Wetlands: Ecology, Conservation and Management

Volume 5

Series Editor
C Max Finlayson
Albury, New South Wales, Australia

The recognition that wetlands provide many values for people and are important foci for conservation worldwide has led to an increasing amount of research and management activity. This has resulted in an increased demand for high quality publications that outline both the value of wetlands and the many management steps necessary to ensure that they are maintained and even restored. Recent research and management activities in support of conservation and sustainable development provide a strong basis for the book series. The series presents current analyses of the many problems afflicting wetlands as well as assessments of their conservation status. Current research is described by leading academics and scientists from the biological and social sciences. Leading practitioners and managers provide analyses based on their vast experience.

The series provides an avenue for describing and explaining the functioning and processes that support the many wonderful and valuable wetland habitats, such as swamps, lagoons and marshes, and their species, such as waterbirds, plants and fish, as well as the most recent research directions. Proposals cover current research, conservation and management issues from around the world and provide the reader with new and relevant perspectives on wetland issues

More information about this series at http://www.springer.com/series/7215

C Max Finlayson • Pierre Horwitz
Philip Weinstein
Editors

Wetlands and Human Health

Springer

Editors
C Max Finlayson
Institute for Land, Water and Society
Charles Sturt University
Albury
New South Wales
Australia

UNESCO-IHE
Institute for Water
Education
Delft
The Netherlands

Philip Weinstein
School of Biological Sciences
The University of Adelaide
Adelaide
South Australia
Australia

Pierre Horwitz
School of Natural Sciences
Edith Cowan University
Joondalup
West Australia
Australia

ISSN 1875-1261   ISSN 1875-127X (electronic)
Wetlands: Ecology, Conservation and Management
ISBN 978-94-024-0373-2   ISBN 978-94-017-9609-5 (eBook)
DOI 10.1007/978-94-017-9609-5

Springer Dordrecht Heidelberg New York London
© Springer Science+Business Media Dordrecht 2015
Softcover reprint of the hardcover 1st edition 2015

Printed on acid-free paper

Springer is part of Springer Science+Business Media (www.springer.com)

Foreword

This book is about wetlands in their broadest sense—as places of water on land, and where water shapes the land. And it is about health in its richest sense—the well-being of people, beyond ill-health or the absence of disease. In some parts of the world these two domains of human thought and existence tend to be separated, into 'environment' and 'health' disciplines and sectors. They only come together when the environment impinges on humans, like an infectious disease, requiring that the environment be manipulated to avoid such possibilities.

We can't continue to see the world, and our relationships with it, like that. In this book we adopt an ecosystem approach to human health, which starts from the premise that Earth systems are foundational for human societies and the well-being of people; the two are interdependent. This is much more like an Indigenous world-view where people, the land and the water are together.

To do this we bring together two concepts—from health promotion we recognise the valuable 'settings' approach, first enunciated in the Ottawa Charter (WHO 1986). The ecological and systems-based perspectives in this approach can be drawn upon to be much more explicit about the ecosystem, in this case the wetland, as the 'setting' in which people "take care of each other, our communities and our natural environment" (see Parkes and Horwitz 2009). The setting includes the institutional and governmental aspects required to deliver health services, to address health inequalities, and to intervene for public health.

From ecological economics we have recognised the 'ecosystem services' approach, which starts from the assumptions that main stream market economics externalises the environment, that we take it for granted, and that we do so at our peril. The ultimate ecosystem service is the availability of water, a chemical compound essential for life. Draining a wetland to avoid mosquito breeding and the transmission of disease, or damming a river to avoid the miserable effects of flooding, at best makes the calculated trade-off to deliver a net benefit; at worst it deprives local people of essential ecosystem services. Either way, the consequences may be more severe than the original problem conceived.

Examining ecosystem services provides analysts with a more effective way to communicate with decision-makers, with the potential for enhancing planning, implementation and assessment efforts for integrated land and water management

(UNU-IHDP 2014). Ecosystem services can also be more clearly linked to aspects of human well-being like health, security, good social relations and basic materials for good life (see Millennium Ecosystem Assessment 2005).

The book draws from the examples given in the Ramsar Convention on Wetlands' technical report *Healthy Wetlands Healthy People* (Horwitz, Finlayson and Weinstein 2012), and goes beyond; the organisation of chapters is more clearly related to the ways in which wetlands as settings contribute to human health. Chapters 3–8 deal with wetlands as playing a critical role in food security and provision of medicinal products, wetlands as sites of exposure to infectious diseases or their vectors and toxic materials, and wetlands as places for livelihood and lifestyle contributions to human well-being. Chapter 9 demonstrates how a wetland can become the setting in which natural disasters are concentrated and their consequences for human well-being are addressed by societal responses and preventative actions. Chapters 10 and 11 follow, firstly by examining the ways in which the public sector in general, and the health sector in particular, might intervene to enhance human well-being by addressing the erosion of ecosystem services in wetlands, and secondly by arguing for specific human health guidance for wetland managers.

References

Horwitz, P., Finlayson, M. and Weinstein, P. (2012). *Healthy wetlands, healthy people. A review of wetlands and human health interactions.* Published jointly by the Ramsar Convention on Wetlands and the World Health Organisation. Ramsar Technical Report No. 6. Gland, Switzerland.

Millennium Ecosystem Assessment (2005). *Ecosystems and Human Well-being: General Synthesis. Millennium Ecosystem Assessment.* Available at: http:// www.unep.org/maweb/en/Synthesis.aspx

Parkes, M. and Horwitz, P. (2009).Water Ecology and Health. Ecosystems as 'settings' for health and sustainability. *Health Promotion International* 24: 94–102.

UNU-IHDP. (2014). Land, Water, and People From Cascading Effects to Integrated Flood and Drought Responses. Summary for Decision-Makers. Bonn: UNU-IHDP.

WHO (1986) Ottawa Charter for Health Promotion. World Health Organisation, Geneva.

C Max Finlayson
Pierre Horwitz
Philip Weinstein

Preface

Wetlands provide a wonderful wealth of ecosystem services and benefits to people, of which many are linked to human health and well-being, including water provision and its purification, resilience to climate change, storm protection, erosion prevention and food provisioning, among others. Wetlands are also crucial to the livelihoods and subsistence of communities, as they are central to the provision of food and water security. For example, wetlands can serve as nurseries for fish, provide crops, such as rice and timber, and recharge groundwater supplies. They can also be sources of spiritual renewal, leisure, tourism and recreation. As a result, degradation of the quality or ecological character of wetlands is often linked to a decrease in these benefits and services to people, and in turn the well-being and health of those who depend upon them.

Wise use of wetlands, which is their maintenance through the implementation of ecosystem approaches, supports sustainable development and is central to the health of wetlands and the health of people. "Healthy people, Healthy wetlands" should be a mantra for our modern world.

Recognizing this important relationship, the Conference of the Contracting Parties of the Ramsar Convention has adopted a number of Resolutions relating to wetlands and human health, such as Resolution X. 13, *the Changwon Declaration on human well-being and wetlands*. The Declaration calls for including the interlinkages between wetlands and human health as key components of national and international policies, plans and strategies in light of the significant health benefits that wetlands provide to people.

The most recently adopted Resolution on this matter is Resolution XI. 12, *Wetlands and health: taking an ecosystem approach*, builds on Ramsar Technical Report No. 6 on *Healthy wetlands, healthy people: A review of wetlands and human health interactions*, and calls for the promotion and delivery of an ecosystem approach to healthy wetlands and their catchments with proposals for integrated methodologies and actions across relevant sectors.

In this light, this book explores the complexities of the relationships between wetlands and humans, through a general treatment of ecology-health issues for the wetlands and public health sectors. Furthermore, this book seeks to further build on the strong message that degrading or converting wetlands, thus stopping them from delivering their services, can have serious adverse effects on human health.

Christopher Briggs
Secretary General,
Ramsar Convention on Wetlands

Contents

ix

Contributors

May Carter School of Natural Sciences, Edith Cowan University, Joondalup, WA, Australia

Scott Carver School of Biological Sciences, University of Tasmania, Hobart, TAS, Australia

Angus Cook School of Population Health, University of Western Australia, Crawley, WA, Australia

Anthony B. Cunningham School of Plant Biology, University of Western Australia, Nedlands, WA Australia

Bonnie T. Derne Queensland Children's Medical Research Institute, Brisbane, Australia, WHO Collaborating, Centre for Children's Health and Environment, The University of Queensland, Brisbane, Australia, Royal Children's Hospital, Herston, QLD Australia

C Max Finlayson Institute for Land, Water and Society, Charles Sturt University, Albury, NSW, Australia
UNESCO-IHE, Institute for Water Education, Delft, The Netherlands

Pierre Horwitz School of Natural Sciences, Edith Cowan University, Joondalup, WA, Australia

Aaron P. Jenkins School of Natural Sciences, Edith Cowan University, Joondalup, WA, Australia

Stacy Jupiter Wildlife Conservation Society, Suva, Fiji

Ritesh Kumar Wetlands International—South Asia, New Delhi, India

Colleen L. Lau Queensland Children's Medical Research Institute, Brisbane, Australia, WHO Collaborating, Centre for Children's Health and Environment,

The University of Queensland, Brisbane, Australia, Royal Children's Hospital, Herston, QLD Australia

Paul T. Leisnham Ecosystem Health and Natural Resource Management, Department of Environmental Science and Technology, University of Maryland, College Park, MD, USA

Matthew P. McCartney International Water Management Institute, Vientiane, Lao PDR

Lisa-Marie Rebelo International Water Management Institute, Vientiane, Lao PDR

Anne Roiko School of Medicine, Griffith University, Southport, QLD, Australia

Sonali Senaratna Sellamuttu International Water Management Institute, Vientiane, Lao PDR

David P. Slaney Institute of Environmental Science and Research Ltd, Wellington, New Zealand and Barbara Hardy Institute, University of South Australia, Adelaide, SA, Australia

Peter Speldewinde Centre of Excellence in Natural Resource Management, The University of Western Australia, Albany, WA Australia

Philip Weinstein School of Biological Sciences, The University of Adelaide, Adelaide, SA, Australia

Wetlands as Settings for Human Health —the Benefits and the Paradox

C Max Finlayson and Pierre Horwitz

Abstract As wetlands provide many valuable ecosystem services and are amongst the most degraded ecosystems globally, further degradation could greatly affect the well-being and health of people dependent on them. Healthy wetlands are generally associated with enhanced ecosystem services and improved outcomes for human health, and unhealthy wetlands with degraded ecosystem services and poor outcomes for human health. However, the relationships can also be paradoxical with some direct benefits for human health leading to the loss of other ecosystem services, in particular regulating and supporting services, and the enhancement of others, leading to poor outcomes for human health. This results in a health paradox whereby there is a loss of regulating and supporting services from steps to enhance human health. Examples of the health paradox include: drainage of wetlands for malaria control; conversion of a wetland into a reservoir to store water for human consumption and irrigation; and regulation of rivers for flood mitigation activities to alleviate loss of life or property. A wetland paradox also occurs when there are poor outcomes for human health as a consequence of the maintenance or enhancement of ecosystem services. Examples of the wetland paradox includes: urban wetlands protected for nature conservation can also support mosquitoes and other vectors, and expose humans to vector-borne diseases; and the maintenance of large woody debris in rivers which slows down water flows, and contributes to the trophic web and is a recreational hazard for swimming or boating. In response a framework for the conceptualisation of human and wetland relationships, including the paradoxical situations has been provided based on the concept of wetlands as settings for human health. This enables the trade-offs that have and will occur between wetland ecosystem services and human health to be addressed.

C. M. Finlayson (✉)
Institute for Land, Water and Society, Charles Sturt University, Albury, Australia
UNESCO-IHE, Institute for Water Education, Delft, The Netherlands
e-mail: mfinlayson@csu.edu.au

P. Horwitz
School of Natural Sciences, Edith Cowan University, Joondalup, WA, Australia
e-mail: p.horwitz@ecu.edu.au

Keywords Wetland settings · Ecosystem services · Health paradox · Wetland paradox · Trade-offs · Biodiversity · Ecological character · Ecosystem health · Agriculture · Water · Livelihoods · Diseases

Introduction

The complexities of the interactions that occur between people and wetlands has been addressed more and more in recent years, for example, through global assessments of water, biodiversity and the wider environment (Falkenmark et al. 2007; Arthurton et al. 2007; Gordon et al. 2010; Armenteras and Finlayson 2011). These assessments, largely following the lead of the Millennium Ecosystem Assessment (MEA 2005), have focused on the benefits that can accrue by promoting the positive relationships that can exist between human well-being and livelihoods, as expressed through the Millennium Development Goals, and wetlands. That is, they have focussed on wetlands as settings for human well-being, including human health and livelihoods, through the provision of ecosystem services, encompassing provisioning, regulating, supporting and cultural services (as defined by the MEA 2005) as well as settings for biodiversity conservation.

At the same time these assessments have provided further documentation that wetlands and wetland-dependent species are in severe decline globally, as are the many ecosystem services that they provide for many people. Given projected increases in the demand for food and fresh water it is expected that wetlands will face increased pressures and further decline in the benefits that they provide for large numbers of people, in particular for those people who depend most directly on wetlands for their sustenance and well-being (Falkenmark et al. 2007; Gordon et al. 2010). Wider recognition that wetlands are important settings for human well-being and for biodiversity conservation is seen as an important step if the decline of biodiversity and ecosystem services from wetlands is to be stopped, let alone reversed (Horwitz and Finlayson 2011). The latter is important—global efforts to reverse the decline of wetlands and wetland species have not kept pace with the rate of decline (MEA 2005; Armenteras and Finlayson 2011). In other words, despite the problems being articulated for several decades (MEA 2005) the responses to the loss of biodiversity and ecosystem services have been inadequate to halt the decline. Further, efforts to ensure greater equity in access to and sharing of the benefits that accrue from biodiversity are increasingly seen as important steps in changing this situation (Armenteras and Finlayson 2011).

The complexities of the relationships between wetlands and people are explored in this book through a general treatment of ecology-health issues for both the wetland and public health sectors, in recognition that both sectors have a vital role to play in ensuring the maintenance of the benefits provided by healthy wetlands. The corollary, namely that disrupting the provision of ecosystem services has adverse impacts on human health, is also examined as a prelude to examining ways in which multi-disciplinary research and practice (including community participation) can be enhanced and policies generated to support ecosystem and human health concurrently.

A key premise behind the abovementioned intent is that the environmental health problems of the twenty-first Century cannot be addressed by the traditional tools of ecologists or epidemiologists working in their respective disciplinary silos; this is clear from the emergence and re-emergence of public health and human well-being problems such as cholera pandemics and mosquito borne disease, as well as the impact of climate change and episodic events and disasters (e.g. hurricanes). The Millennium Ecosystem Assessment concluded that genuine cross-disciplinary approaches were necessary to tackle these problems (MEA 2005), a theme subsequently taken up syntheses provided in the Global Environment Outlook (UNEP 2007, 2011). This book brings the disciplines of ecology and health sciences closer to such a synthesis for researchers, teachers and policy makers interested in or needing information to manage wetlands and their interconnected human health and well-being issues.

While recent global assessments and syntheses provide a basis for many of the technical concepts covering health and wetlands that are expanded in the book, they do not, on the whole, explore the technical knowledge and information that supports the intricacies of the interactions between components of the wetlands and people. Similarly, the major text books covering wetland ecology focus on the science of wetland populations and ecological processes and not on human health issues, responses or specific interactions between wetlands and people. This book brings the disciplines of ecology and health sciences closer together with a synthesis for researchers, teachers and policy makers of the relationships that exist between wetlands and human health—relationships that are fundamental to a sustainable future, but also contains what we have termed a health paradox and a wetland (environment) paradox (Horwitz and Finlayson 2011).

The Human-Wetland Nexus

The relationships between humans and wetland ecosystems go back many millennia with hunter-gatherers being directly dependent on the availability of resources in the immediate environment, foremost being a reliable and clean source of drinking water, but also for food and materials for making tools, shelter, and for fuel for heating and cooking (Junk 2002; Gopal et al. 2008). Over many years and in many places, people developed agriculture, including increasingly intensive use of wetlands for grazing, cropping and horticulture, and eventually changed the manner in which wetlands were managed, including the spread of wide-scale detriment, largely through the expansion of agriculture (Finlayson et al. 2005), but still with a large dependency on the provision of ecosystem services (Falkenmark et al. 2007; Gordon et al. 2010). The continued importance of agriculture in wetlands is evident from analyses of the extent of agriculture in wetlands listed as internationally important under the Ramsar Convention (Ramsar sites); for example, in 2006 some 78 % of Ramsar sites globally were found to support some form of agriculture (Rebelo et al. 2009a), and similarly, in 2008 some 93 % of Ramsar sites in sub-Saharan Africa supported agriculture (Rebelo et al. 2009b).

Unfortunately, the increasing extent of human exploitation and modification of the environment has also adversely affected the health of wetlands, some of which have been lost or degraded to an extent whereby they no longer provide the ecosystem services that previously supported human well-being and health (Revenga et al. 2000; Agardy and Alder 2005; Finlayson and D'Cruz 2005). Sources of drinking and irrigation water have dried, leading to thirst, starvation and population displacement; toxic pollutants have poisoned waters, fish and people; alterations to water regimes and vegetation structures have led to hardship, epidemics, and wide-spread environmental degradation and adverse consequences for people (Horwitz et al. 2012).

On the other hand, changes in land cover and land use to accommodate expanding agriculture and industrial development have had many beneficial outcomes for many people, for example, through increased irrigation and food production. Unfortunately, many agricultural systems have been managed as though they were disconnected from the wider landscape, with scant regard for maintaining the ecological components and processes that underpinned their sustainability (Molden et al. 2007). The consequences of such approaches include the loss of provisioning services such as fisheries, loss of regulating services such as storm protection and nutrient retention, with negative feedback on food and fibre production. Human health has also suffered in a direct sense, for example, through the increased prevalence of insect-borne disease or through changes in diet and nutrition or the loss of regulating services, such as erosion control and the amelioration of floods (Corvalan et al. 2005). People in rural areas who use a variety of ecosystem services directly for their livelihoods are likely to be the most vulnerable to such changes in ecosystems (MEA 2005).

Finlayson et al. (2005) emphasised that failure to tackle the loss and degradation of wetland ecosystems and their species, such as that caused by the development of agriculture and water resources, could undermine progress toward achieving the human health and poverty components of the Millennium Development Goals. The first United Nations World Water Development Report noted that a healthy and unpolluted natural environment was essential for human well-being and sustainable development, and further stressed that wetland (aquatic) ecosystems and their dependent species provided a valuable and irreplaceable resource base that helped to meet a multitude of human and ecosystem needs which are essential for poverty alleviation and socio-economic development (UN-WWAP 2003). The report also noted that human health provided one of the most striking features of the link between water and poverty.

The adverse consequences of increased interactions between people and wetland ecosystems have received more attention in recent years with the Millennium Ecosystem Assessment (MEA 2005) in particular emphasising the strength of the fundamental relationship between ecosystems and human health and poverty, and therefore the importance of developing environmental management strategies that support the maintenance of both wetland health and human health concurrently. Almost in parallel it has become apparent that many environmental health problems cannot be solved by 'traditional' health approaches alone. Rather, broader approaches are needed to analyse interactions between humans and the surrounding environment (Corvalon et al. 2005), often drawing on a wider scientific base, including ecological and social sciences, and accepting that humans are not separable from the complex vagaries of the natural environment.

A fundamental and underlying part of this complexity is the paradox that healthy wetlands (*sensu* Ramsar Wetlands Convention; Finlayson and Weinstein 2008) can provide many valuable ecosystem services as well as support vectors for water-borne diseases (Corvalon et al. 2005). The complexity of such relationships is shown by the historical links between malaria and humans in parts of Europe (O'Sullivan et al. 2008). If wetland health and human health are treated as being inextricably linked it should be no surprise that the incidence of many diseases varies with short- and long-term changes in wetland health. By extension, for a variety of vector-borne, water-borne and other 'environmental' diseases, appropriate, scientifically based public health interventions can only be devised with an understanding of the relationship between wetland health and human health and the ecology of the vectors and diseases. The interactions and reciprocity of the complex interactions between people and wetlands is also illustrated by the debilitating effect of HIV/AIDS which reduces the capacity of groups of people to support their wider well-being through fishing and other basic activities (Mojola 2009).

As wetlands provide many valuable ecosystem services and are amongst the most degraded ecosystems globally, further degradation could greatly affect the well-being and health of people dependent on them both directly and indirectly. In response, the Ramsar Wetland Convention has placed more attention on developing the scientific concepts behind the metaphor 'healthy wetlands, healthy people' and sought more understanding of how people and wetlands interact, for example, through analyses of the interactions between agriculture and wetlands (Falkenmark et al. 2007; Wood and van Halsema 2008) and fisheries and wetlands (Kura et al. 2004), and in this instance, the interactions between human health and wetlands (Horwitz et al. 2012). The metaphor 'healthy wetlands, healthy people' implies an interaction between wetland ecology and management and the health of people with consequent social and cultural interactions between people and wetlands. This is seen as an extension of the multi-disciplinary approaches adopted through the Millennium Ecosystem Assessment (MEA 2005) and subsequent global assessments that have addressed human well-being and ecosystem services (Molden et al. 2007; UN-WWAP 2006; UNEP 2007). The interactions between human health and wetlands are expanded in this book through an examination of the linkages between human health and ecosystem services obtained from wetlands; the emphasis being on human health as a component of human well-being and linked inextricably with wetland health.

Wetlands as Settings for Human Health

With this background the purpose of the book is to review and map out the relationships and issues concerning the wise use of wetland ecosystems and human health, including information and concepts from the Millennium Ecosystem Assessment and its synthesis reports (www.millenniumassessment.org). Specific issues that have been addressed include:

- wetland health and ecological character of wetlands;
- human health and wetland ecosystem services;
- the effects on human health of disruptions to wetland ecosystem services;
- economic values and incentives for supporting human health;
- global trends affecting wetlands and human health; and
- responses and interventions for maintaining the ecological character of wetlands and supporting human health.

In addressing these issues the trade-offs between ecosystem maintenance and the risk of human diseases and ill-health have been considered with comments provided on the complexity of making decisions and choices that support ecosystems and the services that they provide. In doing this the following issues have been addressed—declines in water quantity and quality, including waterborne pollutants; human sanitation; water-related diseases; disease emergence related to small and large dams; increased land use in marginal landscapes leading to closer disease contacts; implications of climate change for human health issues associated with wetlands; human nutrition and wetlands; and wetlands as sources of beneficial drugs. A further section is added on the effect of global trends on wetlands and human health with attention being drawn to the complex interactions with global climate change.

The depth and detail of coverage of the above have benefited by the accessibility of information in recent global overviews such as the Millennium Ecosystem Assessment (MEA 2005), the World Water Development Report (UN-WWAP 2003, 2006), the Comprehensive Assessment of Water Management in Agriculture (Molden 2007), and the Global Environmental Outlook (UNEP 2007, 2011). These overviews represent both a global consensus by scientists on key issues affecting wetland ecosystems, water and people, and up-to-date widely reviewed compilations of science-based evidence. These are particularly important when considering the implications of the achievement of the Millennium Development Goals that may run counter to efforts focussed on wetland conservation with an emphasis on the biodiversity in virtual isolation of wider ecosystem issues. The Millennium Ecosystem Assessment in particular has emphasised the strength of the fundamental relationship between wetland ecosystems and their services and human health, and therefore the importance of developing environmental management strategies that support the maintenance of both wetland health and human health concurrently (Finlayson et al. 2005). It is contended that at a metaphorical level the linkages are being established—further scientific evidence is needed to support these and enable more informed decisions that consider the complexities involved.

The importance of wetlands for humans, in particular in relation to their health and well-being was explored by Horwitz et al. (2012) in a landmark report that considered wetlands as settings that supported and even determined human health and well-being in a number of ways, including the provision of safe water and food and support for livelihoods, but also as places where people could be exposed to pollution, toxicants or infectious diseases. They concluded that wetland settings could "... *either enhance or diminish human health depending on the ecological functioning of wetlands and their ability to provide ecosystem services.*" and that wetland loss and degradation would have consequences for human health, and that adverse outcomes were likely to be distributed unequally, possibly along socio-economic lines.

Horwitz and Finlayson (2011) further explored the concept of wetlands as settings for human health by considering the commonality of issues contained within the concepts of the ecological character of wetlands and wetland ecosystem services. They explored the modern tendency to assess the condition of wetlands and wetland resources separately from human well-being associated with wetlands (considering wetlands in the broad sense of the word as defined by the Ramsar Convention on Wetlands in 1971, to include, e.g., rivers, lakes, marshes, rice fields, coastal areas), and developed the synergies with the Ramsar Convention's concepts of the wise use of wetlands and the maintenance of their ecological character. The Convention responded to the widening gap between wetland conservation and the use of wetlands by people by equating the terms *wise use* and *ecological character* with the maintenance of ecosystem services. This was done by adopting the framework developed by the Millennium Ecosystem Assessment as a framework for the wise use of wetlands (Fig. 1) and updating the definitions of wise use and ecological character (Ramsar Convention Secretariat 2010; Davidson and Finlayson 2007; Finlayson et al. 2011).

> *Ecological character* is defined as "the combination of the ecosystem components, processes, and benefits or services that characterize the wetland at a given point in time.
> *Wise use* of wetlands is defined as "the maintenance of their ecological character, achieved through the implementation of ecosystem approaches, within the context of sustainable development."

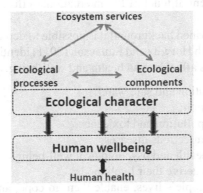

Fig. 1 Conceptualisation of the linkage between ecological character (comprising ecological processes and components and ecosystem services, and their interactions) and human health (comprising human health as one constituent)

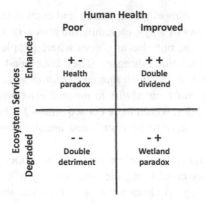

Fig. 2 Categorisation of the settings for wetland and human health based on the condition of ecosystem services and outcomes for human health (adapted from Horwitz and Finlayson 2011). The four settings are: the "*double dividend*" (++) with healthy wetlands and enhanced ecosystem services and improved outcomes for human health; the "*double detriment*" (−−) with unhealthy wetlands and degraded ecosystem services and poor outcomes for human health; the "*health paradox*" (+−) with the loss of regulating and supporting services from steps to enhance human health through; and the "*wetland paradox*" (−+) with poor outcomes for human health as a consequence of the maintenance or enhancement of ecosystem services

The description of the ecological character of a wetland provides a basis for identifying key issues for management, including the role of wetlands in supporting human well-being and health (Fig. 2). It also enabled the relationship between human well-being and health to be expressed pictorially. In making these connections human well-being is treated as a broad notion that includes security, basic materials for a good life, health, good social relations, and freedom of choice and action (MEA 2005), and where *human health* is "a state of complete physical, mental, and social well-being and not merely the absence of disease or infirmity" (WHO 2006). This illustrates how the Ramsar Convention has linked human health with the ecological character of wetlands and given support to the metaphor of "healthy wetlands, healthy people" as a central tenet of the international efforts to make wise use of wetlands globally.

With the above described background it is possible to depict human health issues in a wetland setting, with Horwitz and Finlayson (2011) identifying the following as ways in which wetlands affect human health and well-being. Namely, wetlands are:

- contributors to hydration and safe water;
- contributors to nutrition;
- sites of exposure to pollution and toxicants;
- sites of exposure to infectious diseases;
- settings for mental health and psychological well-being;
- places where people derive their livelihoods;
- places that enrich people's lives, enable them to cope, and allow them to help others;

- places that help absorb the damage of natural disasters; and
- sites where medicinal and other products can be derived.

The relationships that exist between healthy wetlands and the provision of ecosystem services that provide benefits for human health are complicated and include direct or linear links as well as indirect links. Examples of direct links include the provision of food, fuel and fresh water, whereas indirect links include the reduction in vulnerability to extreme events, such as storms and floods, or the amelioration of climate change through carbon sequestration. When wetlands are disrupted these benefits are generally assumed to be reduced or lost; however, the relationship between ecosystem services and human health and wetland health is more complex. In some circumstances degraded wetlands can provide benefits for human health, as shown in the simplified categories outlined in Fig. 2.

The generalised relationships between ecosystem services and human health are outlined in Fig. 2. Healthy wetlands are generally associated with enhanced ecosystem services and improved outcomes for human health (the $++$ or double dividend scenario), and unhealthy wetlands with degraded ecosystem services and poor outcomes for human health (the $--$ or double detriment scenario). However, given the multiplicity of ecosystem services and outcomes for human health, the relationships can also be paradoxical with some direct benefits for human health leading to the loss of [other] ecosystem services, in particular regulating and supporting services, and the enhancement of others, for example, nature conservation in particular environments, leading to poor outcomes for human health. These situations lead to what is described as the "health paradox" (the $+-$ scenario) and the "wetland paradox" (the $-+$ scenario) in Fig. 2.

The health paradox occurs when there is a loss of regulating and supporting services from steps to enhance human health through, for example: (i) drainage of wetlands for malaria control; (ii) conversion of a wetland into a reservoir to store water for human consumption and irrigation; and (iii) regulation of rivers for flood mitigation activities to alleviate loss of life or property. The wetland paradox occurs when there are poor outcomes for human health as a consequence of the mainte nance or enhancement of ecosystem services, for example: (i) urban wetlands protected for nature conservation can also support mosquitoes and other vectors, and expose humans to vector-borne diseases; and (ii) the maintenance of large woody debris in rivers which slows down water flows, and contributes to the trophic web and is a recreational hazard for swimming or boating.

While the scenarios in Fig. 2 simplify the interactions between human and wetland health they do provide a framework for considering the general relationships or settings for these relationships. Horwitz and Finlayson (2011) explain that the simplification is to some extent inevitable given that in any wetland some ecosystem services will be maintained, some embellished and some degraded, and similarly, there is the ever present likelihood that there will be both poor and beneficial health outcomes for various people. The multiplicity of wetland and human health outcomes can seem complex but through careful consideration many can be seen to be causally linked, layered, displaced in space and time, dependent on similar or even the same modifying forces, and form chains of events and outcomes.

The chains of events and outcomes imply that trade-offs for particular aspects of human health will occur when wetlands are modified by human activities that promote or favour one or a few ecosystem services over others. In saying this we note that a comprehensive and specific assessment of how particular wetland ecosystem services, or combinations of services, affect human health has not been undertaken.

Horwitz and Finlayson (2011) have provided a framework for collecting further information and teasing apart the relationships between wetlands and human health. This was developed by providing a joined-up account and by adopting constructs from ecosystem management, ecological economics, public health, epidemiology, and health promotion. They also drew upon an accepted global framework for describing a wetland's ecological character and described how it should include the services that wetlands provide to human welfare. In doing this they made a case for bringing wetland ecosystems to the foreground as the settings and context in which health determinants can be addressed.

The complexities of interactions between wetlands and human health are contained within the conceptualisation of wetlands as settings for human health, including those of a paradoxical nature. The framework for the conceptualisation does not ignore the paradoxes—rather it enables them to be highlighted along with the double dividend and double detriment scenario. This includes what have been termed the "health and wetland paradoxes": (i) the health paradox occurring when there is a loss of regulating and supporting services from steps to enhance human health, and (ii) the wetland paradox occurring when there are poor outcomes for human health as a consequence of the maintenance or enhancement of specific ecosystem services.

Establishing the trade-offs that have and will occur between multiple measures of wetland ecosystem services and multiple measures of human health can be done using the concept of wetlands as settings for human health. The alternative of treating them separately, as has been done in many societies for decades (or longer), may drive us further towards the double detriment rather than towards the double dividend scenario. Looking at wetlands and human health in this manner also enables a further statement about the situation whereby wetlands are seen as valuable, and yet are highly degraded by human activities. The settings concept raises the hypothesis that insufficient attention has been given to the dividends for human health that can accrue from a healthy wetland, and conversely, that more attention has been given to the negative outcomes that can accrue from an unhealthy wetland. Further, it may also be hypothesised that the consequences of the health paradox has received more attention than the consequences of the wetland paradox.

The settings construction enables the complexities of the interactions between wetlands and human health to be explored in a manner than extends far beyond the oversimplification of statements such as "healthy wetlands healthy people". The construction of artificial wetlands in urban environments is an example where wetland settings are considered to bring multiple benefits to people, including, in places, the treatment of wastewater and mediation of flood flows while providing amenity value for people. In many instances however, artificial wetlands are constructed for single purposes, with Everard et al. (2012) describing them as nature without imagination. The restoration of wetlands, as actively promoted by the Ramsar Convention (Alexander et al. 2011; Alexander and McInnes 2012), provides another opportunity to develop the benefits of wetlands as settings for human health and well-being. Recent attention to the creation or restoration of urban wetlands is an area where the concept of wetlands as settings with potentially paradoxical outcomes is most advanced.

The complexities of human interactions with wetlands in urban areas are being explored with increasing attention to the benefits and problems that may arise when wetlands are created, highly modified or restored (McInnes 2014). The same viewpoint may not be as prevalent in more rural settings with smaller populations, such as in the Murray-Darling Basin in south-eastern Australia, where steps to restore the riverine environment have focussed largely on engineering solutions based on hydrological criteria with little consideration of wider values and benefits for people (Pittock and Finlayson 2011). The conceptualisation of wetlands as settings for human health and well-being, taking into account the double dividend and paradox, as described above, is seen as a way of exploring the benefits that can accrue for people from wetlands.

References

Agardy T, Alder J (2005) Coastal systems. In: Millennium ecosystem assessment, vol 1. Ecosystems and human well-being: current state and trends. Findings of the Conditions and Trends Working Group. Island Press, Washington, DC

Alexander S, McInnes RJ (2012) The benefits of wetland restoration. Ramsar Scientific and technical briefing note No. 4. Gland: Ramsar Convention Secretariat

Alexander S, Davidson N, Erwin K, Finlayson MC, McInnes RJ (2011) Restoring our wetlands: an important tool for improving ecosystem services and enhancing human well-being. In: Abstracts of posters presented at the 15th meeting of the subsidiary body on scientific, technical and technological advice of the convention on biological diversity, 7–11 November 2011, Montreal, Canada. Technical Series No. 62. Montreal, SCBD, pp 15–17

Armenteras D, Finlayson CM (2011) Biodiversity. In: UNEP. Keeping track of our changing environment: from Rio to Rio+20 (1992–2012). Division of Early Warning and Assessment (DEWA), United Nations Environment Programme (UNEP), Nairobi

Arthurton R, Barker S, Rast W, Huber M, Alder J, Chilton J, Gaddis E, Pietersen K, Zöckler C, Al-Droubi A, Dyhr-Nielsen M, Finlayson M, Fortnam M, Kirk E, Heileman S, Rieu-Clark A, Schäfer M, Snoussi M, Danling TL, Tharme R, Vadas R, Wagner G (2007) Water. In: United Nations Environment Program. Global Environment Outlook 4-Environment for Development, UNEP, Nairobi, pp 115–156

Corvalan C, Hales S, McMichael A (2005) Ecosystems and human well-being: health synthesis. World Health Organization, Geneva

Davidson NC, Finlayson CM (2007) Developing tools for wetland management: inventory, assessment and monitoring-gaps and the application of satellite-based radar. Aquat Conserv Mar Freshw Ecosyst 17:219–228

Everard M, Harrington R, McInnes RJ (2012) Facilitating implementation of landscape-scale water management: the integrated constructed wetland concept. Ecosyst Serv 2:27–37

Falkenmark M, Finlayson CM, Gordon L (coordinating lead authors) (2007) Agriculture, water, and ecosystems: avoiding the costs of going too far. In: Molden D (ed) Water for food, water for life: a comprehensive assessment of water management in agriculture. Earthscan, London pp 234–277

Finlayson CM, D'Cruz R (2005) Inland water systems. In: Millennium ecosystem assessment, vol 2. Conditions and trends. Island Press, Washington, DC, pp 551–583

Finlayson CM, Weinstein P (2008) Wetlands, health and sustainable development-global challenges and opportunities. In: Ounsted M, Madgwick J (eds) Healthy wetlands, healthy people. Wetlands International, Wageningen, The Netherlands, pp 23–40

Finlayson CM, D'Cruz R, Davidson NC (2005) Ecosystems and human well-being: wetlands and water synthesis. World Resources Institute, Washington, DC

Finlayson CM, Davidson N, Pritchard D, Milton GR, MacKay H (2011) The Ramsar Convention and ecosystem-based approaches to the wise use and sustainable development of wetlands. J Int Wildl Law Policy 14:176–198

Gopal B, Junk W, Finlayson CM, Breen CM (2008) Present state and future of tropical wetlands. In: Polunin NVC (ed) Aquatic ecosystems: trends and global prospects, Cambridge University Press, Cambridge, pp 141–154

Gordon L, Finlayson CM, Falkenmark M (2010) Managing water in agriculture to deal with tradeoffs and find synergies among food production and other ecosystem services. Agric Water Manag 97:512–519

Horwitz P, Finlayson CM (2011) Wetlands as settings: ecosystem services and health impact assessment for wetland and water resource management. BioScience 61:678–688

Horwitz P, Finlayson M, Weinstein P (2012) Healthy wetlands, healthy people: a review of wetlands and human health interactions. Ramsar Technical Report No. 6. Secretariat of the Ramsar Convention on Wetlands, Gland, Switzerland, & The World Health Organization, Geneva, Switzerland

Junk W (2002) Long-term environmental trends and the future of tropical wetlands. Environ Conserv 29(4):414–435

Kura Y, Revenga C, Hoshino E, Mock G (2004) Fishing for answers. World Resources Institute, Washington, DC

McInnes RJ 2014. Recognising wetland ecosystem services within urban case studies. *Marine and Freshwater Research* 65: 575-588.

MEA (Millennium Ecosystem Assessment) (2005) Ecosystems and Human Well-being: synthesis. Island Press, Washington, DC

Mojola SA (2009) Fishing in dangerous waters: ecology, gender and economy in HIV risk. (Population Program Working Paper). Institute of Behavioural Studies, University of Colorado, Colorado

Molden D (ed) (2007) Water for food, water for life: a comprehensive assessment of water management in agriculture. Earthscan, London

Molden D, Frenken K, Barker R, de Fraiture C, Mati B, Svendsen M, Sadoff, C, Finlayson CM (2007) Trends in water and agricultural development. In: Molden D (ed) Water for food, water for life: a comprehensive assessment of water management in agriculture. Earthscan, London, pp 57–89

O'Sullivan L, Jardine A, Cook A, Weinstein P (2008) Deforestation, mosquitoes, and ancient Rome: lessons for today. BioScience 58:756–760

Pittock J, Finlayson CM (2011) Australia's Murray Darling Basin: freshwater ecosystem conservation options in an era of climate change. Mar Freshw Res 62:232–243

Ramsar Convention Secretariat (2010) Ramsar handbooks for the wise use of wetlands. In: Pritchard D (ed) Handbook 1, Concepts and approaches for the wise use of wetlands, 4th edn.

Rebelo L-M, Finlayson CM, Nagabhatla N (2009a) Remote sensing and GIS for wetland inventory, mapping and change analysis. J Environ Manag 90:2144–2153

Rebelo L-M, McCartney MP, Finlayson CM (2009b) Wetlands of Sub-Saharan Africa: distribution and contribution of agriculture to livelihoods. Wetl Ecol Manag 8:557–572

Revenga C, Brunner J, Henninger N, Kassem K, Payne R (2000) Pilot analysis of global ecosystems: freshwater ecosystems. World Resources Institute, Washington, DC

UNEP (United Nations Environment Program) (2007) Global Environment Outlook 4- Environment for Development, UNEP, Nairobi

UNEP (United Nations Environment Programme) (2011) Global Environment Outlook 5- Environment for the future we want. UNEP, Nairobi

UN-WWAP (United Nations & World Water Assessment Programme) (2003) UN world water development report: water for people, water for life. UNESCO and Berghahn Books, Paris and New York

UN-WWAP (United Nations & World Water Assessment Programme) (2006) UN world water development report: water, a shared responsibility. UNESCO and Berghahn Books. Paris and New York

WHO (2006) Constitution of the World Health Organization. WHO. http://www.who.int/governance/eb/who_constitution_en.pdf (17 January 2015)

Wood A, van Halsema GE (eds) (2008) Scoping agriculture -wetland interaction: towards a sustainable multiple-response strategy. FAO Water Report 33, Rome

Public Health Perspectives on Water Systems and Ecology

Angus Cook and Peter Speldewinde

Abstract Human health is directly linked to ecosystem functioning, including the dynamics within wetlands. Wetlands can act to initiate or mitigate biological and chemical contamination. In order to assess the potential health risk, investigations focus on potential exposure pathways and evidence that exposure events are actually linked to adverse health effects. This chapter provides a brief introduction to the principals of health risk assessment in relation to wetlands.

Keywords Public health · Human health · Health risk assessment · Epidemiology · Surveillance · Water-borne disease · Water-borne exposures · Contamination · Disability adjusted life years (DALYs) · Cholera · Human pathogens · Toxic substances · Wetland settings · Ecosystem services

Introduction

Protection of public health requires the maintenance of adequate quality, quantity, accessibility and continuity of water supplies for communities (World Health Organization 2006a). For this reason, the World Health Organisation (WHO) has stressed the importance of "continuous and vigilant public health assessment and review of the safety and acceptability of drinking-water supplies" (World Health Organization 1976). Although considerable effort and investment has been directed towards achieving this health objective, water-borne diseases—that is, those arising from pathogens or chemical contaminants transmitted through water supplies—remain a major cause of global disease (Tebbutt 1998). Over 1.1 billion people globally continue to consume unsafe drinking water (Rodgers et al. 2004; WHO

A. Cook (✉)
School of Population Health, University of Western Australia, Crawley, WA, Australia
e-mail: angus.cook@uwa.edu.au

P. Speldewinde
Centre of Excellence in Natural Resource Management,
The University of Western Australia, Albany, Australia
e-mail: peter.speldewinde@uwa.edu.au

© Springer Science+Business Media Dordrecht 2015
C. M. Finlayson et al. (eds.), *Wetlands and Human Health,* Wetlands: Ecology, Conservation and Management 5, DOI 10.1007/978-94-017-9609-5_2

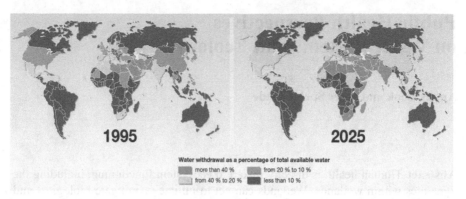

Fig. 1 Patterns of global water withdrawal (http://www.grida.no/publications/vg/water2/page/3289.aspx)

2004a) and a large proportion (88%) of diarrhoeal illnesses has been attributed to unsafe water, sanitation and hygiene (WHO 2004b). In developing countries, diarrhoeal illnesses are ranked third in terms of the total burden of disease (with only HIV/AIDS and lower respiratory infections ranking higher) (WHO 2004a). Approximately 90% of this disease burden occurs in children under the age of 5 years (Pruss et al. 2002) where severe malnutrition and lack of medical care compound the severity of the problem (Leclerc et al. 2002).

In recent decades, issues of global water availability and sustainability have become ever more pressing (Vörösmarty et al. 2010) (Fig. 1). The amount and quality of fresh water is decreasing, and aquifers, rivers and oceans are steadily becoming more degraded and depleted (Fig. 2 and 3). The construction of wells, dams, drains, canals and systems for irrigation have accelerated over the past century and contributed to serious disruptions in flow. Water contamination with chemicals or with infectious organisms is a significant cause of human disease globally.

In response to these growing challenges, there has also been a widening in the scope and activities associated with public health. It has become more generally appreciated that human well-being must be conceptualised within a wider set of environmental processes. The future of our health ultimately depends on the interaction between our species and surrounding physical, chemical and biological environments. These examples of human reliance on the environment can be termed as ecosystem services, where an ecosystem service is defined as the benefits that humans obtain from ecosystems (Corvalán et al. 2005). Four types of ecosystem services are identified: supporting services (such as nutrient cycling), provisioning (such as production of food and fuel), regulating services (such as regulation of climate or disease) and cultural services (which includes recreational and educational uses) (Corvalán et al. 2005). In the context of wetlands, ecosystem services such as provision of drinking water, flood control, source of fuel for water boiling, water for agriculture, can have a dramatic impact on the health of human populations (Horwitz et al. 2012). (See Box 1) Wetlands provide all four ecosystem services (Fig. 4). Our relationship with the biotic world and biogeochemical cycles is in a

Fig. 2 Example of salinisation of river in Western Australia showing degradation of vegetation (Photo Peter Speldewinde)

Fig. 3 Over grazing by cattle causing degradation of waterbody (Photo Paul Close)

Fig. 4 Types of ecosystem
services with examples of
services provided. (Modified
from the MEA Health Syn-
thesis; Corvalán et al. 2005)

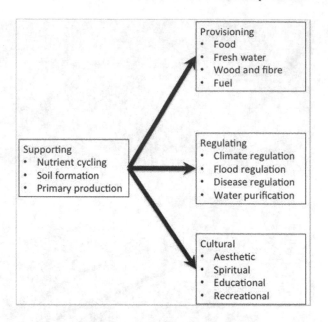

state of permanent flux that is made all the more variable by human interventions. These ideas have been formalised in various ecosystem and health approaches (Patz 2007; Aron and Patz 2001), which place the linkages between the environment and human well-being at their core. In the case of wetlands, three main health requirements are important: access to sufficient and safe water, provision of nutrition and social benefits (Horwitz et al. 2012). These can have direct health impacts (such as floods), ecosystem-mediated impacts (such as reduced food yields) or indirect health impacts (such as population displacement) (Corvalán et al. 2005).

Box 1: Cholera an Example of Ecosystem Services

Cholera is a water-borne disease caused by the bacterium *Vibrio cholera*. It is a disease which has strong environmental drivers which may be examined using the paradigm of ecosystem services. For example, high nutrient levels in the water combined with warm temperatures increases the levels of cholera in the water (regulating services) (Cottingham et al. 2003).

The chapter will provide a brief summary of the relationship between public (community) health and the water cycle (see Boxes 2 & 3). In the first section, a general overview of how water systems and ecologies relate to public health is provided, followed by a systematic approach for assessing health risks from water contaminants. The last two sections will examine how two vital public health activities—the application of epidemiological methods and the use of systems of surveillance—relate to water sources and supplies.

Box 2: Ecosystem and Human Pathogens

Land clearing in South East Asia has resulted in an increase in the human population in areas previously unpopulated or sparsely populated, an increase in potential breeding sites for mosquitoes and a reduction in biodiversity with reduced abundance of mosquito predators. With acceleration of the rate of land clearing there has been an increase in the number of reported malaria cases (Norris 2004). The reduction of biodiversity can lead to an increase in a range of other vector borne diseases, such as Lyme disease, Chagas disease and West Nile Encephalitis (Ostfeld and LoGiudice 2003). In environments with high species diversity, there are a greater proportion of incompetent hosts which attract vectors away from the most competent hosts therefore reducing the prevalence of the disease. Conversely, reduction of biodiversity can lead to a reduction in incompetent hosts and therefore an increase in the disease prevalence (Ostfeld and Keesing 2000).

Box 3: Ecosystems and Recycling of Toxic Substances

The drying of the Aral Sea, due to the diversion of water from the sea for irrigation, and the over use of fertilisers and pesticides (combined with a declining regional economy), has been linked to increased rates of cancers, respiratory conditions, tuberculosis and infant mortality in a number of Central Asian states (Zetterstrom 1998; O'Hara et al. 2000; Crighton et al. 2003). These disorders have been attributed to the increase in airborne dust, a consequence of the drying of the sea, which is laden with residual pesticides and fertilisers.

Water Systems From a Public Health Perspective

Sources and Contamination Processes Maintaining water quality standards is one of the core functions of public health. The two principal forms of water that are available for human use are surface water and ground water. Surface water is the term used to describe any water body that stands or flows above ground, such as lakes, rivers, streams, and so-called impounded water, such as reservoirs and dams. The quality of surface water is sensitive to both the abiotic environment and human/animal activity. Microbial deterioration of surface water quality may result from discharging effluent, wastewater or stormwater into sources waters, faecal contamination of the water from nearby livestock and local fauna, or by humans that utilise surface water and surrounding catchment areas for recreational purposes. Microbial deterioration of water bodies can also occur due to the removal of ecosystem components, which would under normal circumstances reduce the levels of microbial contamination, such as the removal of riparian vegetation which filter nutrients and pathogens prior to surface run off entering water bodies (Barling and Moore 1994).

Groundwater for human use can be extracted from fully saturated soils and rocks through boreholes and wells, or it can be collected through natural outlets, such as springs (Percival et al. 2000). Soil layers provide a barrier to microbial contamination by filtering rain and surface water (Tebbutt 1998) and by enhancing pathogen die-off as water filters through to underlying aquifers. Despite the protection of groundwater because of its relative isolation, it is still possible for microbial and chemical contaminants to reach these sources. Contaminants enter groundwater through direct injection into wells, the leaching of soluble solids or liquids sprayed on the surface (e.g. slurry), leaking or broken sewer lines, seepage from waste reservoirs and landfills, septic tank effluents and infiltration of polluted surface water. Contamination of groundwater can also occur through indirect processes, such as the case of dryland salinity in Western Australia. Removal of deep rooted perennial vegetation has resulted in the rise of the water table which dissolves salts in the soil profile bringing them to the surface resulting in increased salinity of the water. This increase in salinity has been shown to have human health consequences such as increased rates of depression and co-morbid conditions (Speldewinde et al. 2009, 2011), and increased levels of Ross River Virus in the environment (Jardine et al. 2008).

In broad terms, contaminants may be classified according to whether they originate from *point* or *non-point* sources. Point sources are discrete locations—such as effluent outfalls—that release pollutants, whereas non-point sources are more diffuse (such as agricultural runoff) (Hodgson 2004). Agricultural practices are a major source of water contamination, including major human pathogens such as *Cryptosporidium*, *Giardia*, and Campylobacter. Wastes may be carried as direct runoff into surface waters or may collect in impoundments and thereby infiltrate groundwater. A range of manufacturing and industrial wastes and by-products, as well as commercial products, may be discharged into water systems. Well-documented examples include plasticisers and heat stabilisers, biocides, epoxy resins, bleaching chemicals and by-products, solvents, degreasers, dyes, chelating agents, polymers, polyaromatic hydrocarbons, polychlorinated biphenyls and phthalates.

Urban runoff contains a complex mixture of microbiological and chemical contaminants (Gray and Becker 2002). For example, large quantities of hydrocarbons are emitted into the atmosphere and can be washed into water systems. Stormwater (primarily rain from roofs, roads and other surfaces that passes into the drainage system) often contains various debris, animal faeces, oils/grease, soil, metals from road surfaces, pesticides and fertilisers from roadsides.

The Burden of Water-Borne Disease The major microbiological risks arising from human contact with water sources are infection from viruses, bacteria, protozoa and helminths. Human microbial pathogens found in water are often enteric in origin. Important water-borne agents in industrialised countries include the bacteria *Campylobacter*, *Salmonella*, *Shigella* and *E.coli,* the parasites (protozoa) *Crypto-*

sporidium and *Giardia* and the viral agents Hepatitis A, Norwalk virus and rotaviruses. Enteric pathogens enter the environment in the faeces of infected hosts and can enter water either directly through defecation into water, contamination with sewage effluent, or from run-off from soil and other land surfaces. For example, in 1998 *Cryptosporidium* and *Giardia* cysts were found in the water supply of Sydney, Australia. The occurrence of the pathogens in the water supply was attributed to the presence of cattle within the catchment combined with unseasonal high stream flow (Hawkins et al. 2000).

The dynamics of water-borne disease transmission are highly complex and dependent upon multiple variables. One crucial factor is that of pathogen persistence in water. The pathogen must be able to resist environmental stresses and maintain viability while water-borne; otherwise the ability to infect new hosts becomes greatly diminished (WHO 2004b). Persistence in water also relates to temperature, ultraviolet radiation, nutrient availability, chlorine concentration (WHO 2004b), sedimentation, predation, dilution, pH, and the magnitude of pollution loading from discharges and their degree of treatment (McFeters 1990; Olson and Hurst 1990). Human intervention in ecosystems can impact on these stresses, for example clearing of riparian vegetation removes shade and therefore increases water temperature, while increasing the amount of ultraviolet radiation and amount of nutrients entering the water body through surface water runoff. Because of the complicated interactions between the pathogen, environment and the human host, water-borne pathogen ecology remains poorly understood for many microorganisms.

The greatest threats to human health are posed by pathogens with high infectivity, those which can survive or proliferate in water, or those which are resistant to decay or degradation. A number of pathogens remain highly stable through the formation of cysts or spores. For example, *Cryptosporidium parvum* typically has considerable potential to cause gastroenteritis (Havelaar and Melse 2003), given that it is reasonably infective and resistant to chlorination (National Water Quality Management Strategy 2006). *Cryptosporidium* oocysts are widely distributed in source waters across the globe, with oocyst concentration typically high in areas of poor water quality (e.g. areas located near agricultural activity) (Hansen and Ongerth 1991; Lechevallier et al. 1991).

Through the introduction of water sanitation procedures, waterborne diseases in industrialised countries have continually declined since the mid-nineteenth century (Herwaldt et al. 1991; Fewtrell and Bartram 2001). However, more recently, several industrialised countries have noted increases in the number of waterborne disease outbreaks (Lee et al. 2002a). Numerous reasons for these increases have been identified, including water system failures (Moore et al. 1993; Kramer et al. 1996; Levy et al. 1998; Barwick et al. 2000; Lee et al. 2002b), decaying water treatment and supply infrastructure, and the emergence (or improved detection) of new, previously unrecognised, resistant or more virulent organisms (Fewtrell and Bartram 2001).

There has been a growing international focus on the development of guidelines or regulations for chemical parameters in water systems. Many chemicals are in low concentrations, but pose a potential risk from chronic exposures or bioaccumulation. For example, pesticides commonly enter water systems from agricultural activities, via influxes of stormwater, or direct disposal by households or commercial premises. During the last few decades, there has been growing evidence of hormonally-related abnormalities in a wide range of species (Matthiessen 2003). These have included invertebrates (Oehlmann and Schulte-Oehlmann 2003), fish (Jobling and Tyler 2003), aquatic mammals (Fossi and Marsili 2003), reptiles (Guillette et al. 2007; Guillette and Iguchi 2003) and birds (Giesy et al. 2003). Chemical contaminants are believed to be responsible for many of these abnormalities, acting via mechanisms leading to alteration in endocrine function. This phenomenon, known generally as 'endocrine disruption', has been identified by the World Health Organisation as an issue of global concern (World Health Organization 2005). The chemicals implicated have been collectively termed 'endocrine disrupting chemicals' (EDCs), or simply 'endocrine disruptors'. Pharmaceutically-active compounds (PhACs) are another group of compounds that have led to concerns about adverse health outcomes (National Water Quality Management Strategy 2008; Toze 2006; Kolpin et al. 2002). As well as directly impacting on human health through direct ingestion these chemicals can have indirect impacts on human health through ecosystem services, such as decreases in fish catches.

Principles of Health Risk Assessment

Risk is defined by The National Water Quality Management Strategy (National Water Quality Management Strategy 2006) as "The likelihood of a hazard causing harm in exposed populations in a specified time frame, including the magnitude of that harm." In general, any assessment of risk to community health seeks to estimate the potential impact of an activity or process on a specified human population within a specified time period (in the past, now and/or in the future). There is usually an emphasis on evaluating factors in the environment that impact on disease in order to make prevention of disease possible.

In general terms, health risk assessments may be conceptualised as an investigation of (i) *exposure pathways:* whether there are multiple and/or interacting hazards (e.g. from various sources); the routes of exposure; and projected contaminant intakes in at-risk populations; coupled with (ii) *evidence on the progression to health end-points:* for example, toxicological analyses (such as animal/*in vitro* studies) and/or epidemiology analyses (studies in human populations) may be used to assess whether the hazards are likely to produce any adverse health effects, to explore the relationship between toxicant 'dose' and occurrence of a particular disease, and to calculate rates of disease in the given population.

The formal methods for assessing risks arising from water contaminants include:

- Epidemiological investigations (refer to 3.4 below)
- Qualitative risk assessment (with risk ranking): Qualitative assessments usually combine an indication of the likelihood of identified hazards causing harm (in exposed populations or receiving environments in a specified timeframe) with the severity of the consequences characterised using a categorical scale (e.g. low; moderate; high; very high).

- Quantitative Microbial Risk Assessment (QMRA) (National Research Council (U.S.). Committee to Evaluate the Viability of Augmenting Potable Water Supplies with Reclaimed Water 1998; EnHealth Council 2002): QMRA is a mathematical risk assessment model that can accurately predict the risk associated with exposure to pathogens (mainly bacteria, viruses and protozoa) in water sources (Haas et al. 1999; Havelaar and Melse 2003). A full assessment progresses through a number of stages: Hazard assessment; Exposure assessment; Dose-response analysis; Risk characterisation.
- Risk assessment for chemical contaminants progresses through similar stages to those for microorganisms (European Commission Brussels 2003). Hazard assessment investigates the inherent properties of chemicals by collecting and comparing relevant data on, for example, physical state, volatility and mobility as well as potential for degradation, bioaccumulation and toxicity. Models are used to assess the distribution of contaminants in the environment (soil, water, air) and in tissue (animals, humans). Estimations of human health risks from exposure to specific chemicals are generally based on extrapolations of the results of toxicological experiments on animals. These extrapolations provide standard human 'dose-response' relationships for the chemicals. The validity of the data and the weight-of-evidence of various toxicity data are assessed. For example, the International Agency for Research on Cancer (IARC) grades hazards according to various likelihoods of whether they are carcinogenic.

In the field of water and heath, the risk assessment models described above have been extended using the Stockholm Framework (World Health Organization 2006c), which incorporates the concept of *disability adjusted life years (DALYs)* to assess health outcomes from different disease exposure routes (Fig. 5). The DALY is a metric that considers health burden in terms of years of life lost (due to premature death) and years "lost" to disability, with different "weights" assigned to medical conditions depending on their severity. When deriving DALYs for individual hazards, both acute public health effects (such as diarrhoeal disease and even death) and chronic public health effects (such as cancer) are considered (National water quality management strategy 2006). DALYs have been used extensively by agencies such as the World Health Organization (WHO) to assess disease burdens and to identify intervention priorities associated with a broad range of environmental hazards (WHO 2004b).

The information gathered from risk assessment help to inform the next stage: *risk management,* or the process of evaluating possible action and alternatives and then implementing these in response to the risk assessment. This requires careful consideration of the options and strategies to reduce risk taking into account all factors including practicality, social and political implications. A common intervention is exposure control, in which the hazard of concern is controlled at source or at some point in the exposure pathway.

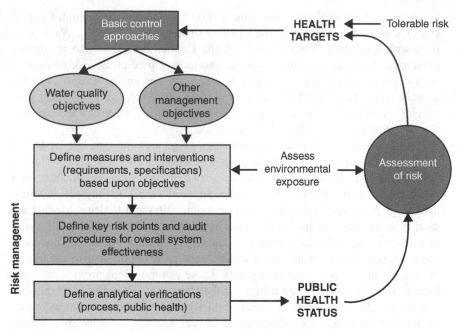

Fig. 5 Outline of Stockholm Framework for risk assessment. (World Health Organization 2006b)

Principles of Epidemiology in Relation to Water-Borne Exposures

Common Measures Used in Epidemiology Epidemiology is the study of disease patterns in populations. In a public health context, there is often an emphasis on discovering the causes, determinants or risk factors of a disease in order to make prevention possible (Gordis 2009; Rothman et al. 2008; Szklo and Nieto 2007). There are many epidemiological study designs and techniques, but the key questions usually relate to: *how and when do we measure the exposure and health effect,* and *which populations should we select and compare?* Analyses are conducted (in an attempt) to indicate the degree of risk that an individual (or population) with a particular exposure pattern, lifestyle, genetic profile or other determinant has of contracting a specific disease.

Epidemiologists often report their findings as the *magnitude of association* between exposure and disease: these estimates may be based on *absolute differences* (obtained by subtracting disease rates in the unexposed from rates in the exposed) or *relative differences or ratios* (obtained by dividing disease rates in the exposed by rates in the unexposed) (Hennekens et al. 1987). A common measure of risk is the relative risk, which indicates how many times more likely "exposed" persons (such as residents of a wetland with high levels of nutrient contamination which cause increased abundance of mosquitoes) are to develop the disease (such as malaria), relative to "non-exposed" persons (such as residents of a non-contaminated wetland).

No single epidemiological measure can describe adequately the health of a community. Only through using a number of indices can a reasonably complete picture of a population's health be constructed. One of the most easily measured indicators is *mortality rate,* or the number of deaths in a population over a set period (usually a year). Frequency of disease is another commonly employed measure, such as the *incidence rate* (the number of new cases occurring in a given population per unit time). The *prevalence proportion* of disease is the proportion of people who are ill with the disease within a given population at a given point in time, accounting for both new and existing cases with the disease. It is often said that the prevalence reflects the *current burden of disease* in a population or community. Many of these measures are limited by a predominantly physical focus: the psychological, social, and functional aspects of poor health and disability are of great importance to individuals but are difficult to measure.

The Role of Outbreak Investigations National public health authorities may undertake or direct research to evaluate the role of water as a risk factor in disease. Of the various epidemiological tools that may be applied to water and health, it is the *outbreak investigation* that often provides useful information regarding which failures in the water supply and distribution chain are associated with risks to public health. Outbreak investigations also provide information on non-water exposure pathways (e.g. food) that could be related to the suspect pathogen (or chemical agent). (Organisation for Economic Co-operation and Development and World Health Organization 2003)

Outbreak investigations ask a range of crucial questions about how the disease in question is affecting the population. These include defining the *"time, place, person"* descriptors of the disease of interest. The "time" component of an outbreak investigation seeks to determine: *when and how often does the disease occur?* and, *is the frequency different now than in the past?* The time component may encompass long term trends (changes in the incidence of disease over a number of years or decades), cyclical or seasonal trends (which have some kind of recurring pattern; for example, gastrointestinal disease is more common is summer when the temperature and humidity favours the multiplication of enteric pathogens); or short term trends (temporary aberrations in the incidence of disease).

The "place" component of an outbreak investigation seeks to determine: *where are the rates of the disease highest and lowest?* Water-borne diseases may vary with: biological environment (including the local ecology, such as presence of animal reservoirs for disease); physical environment (such as climate and other abiotic factors affecting the capacity of pathogens to survive); and the sociocultural environment (sanitation and hygiene practices; cultural practices). Mapping and geographic information systems (GIS) may be used to display or investigate some attribute, theme, or other variable, such as the incidence rate of a disease in a particular area. In water-borne diseases, GIS may be used to define the water distribution area, to assess the exposed population, and to evaluate possible relationships between the water supply and disease (Hunter et al. 2003).

The "person" component seeks to determine: *who is getting the disease?* Important factors in infectious disease susceptibility include age, gender, ethnicity, occupation, personal habits and co-existing disease. In water-borne disease, there may be susceptible subgroups or those with characteristic exposure patterns (e.g. private water supplies).

Sporadic outbreaks may be investigated in greater detail (Hunter et al. 2003). Such events may lead to full review of source and treatment of the water supply, area of distribution, recent changes in treatment or supply. Often subtyping may be performed on laboratory samples to detect possible temporal or spatial clusters. Interpretation of these and other data are used to institute control measures to manage the outbreak. This stage usually requires the treatment of cases, and/or the introduction of control measures to reduce the spread of the epidemic or prevent its recurrence.

Principles of Disease Surveillance in Relation to Water-Borne Exposures

In a public health sense, surveillance is viewed as "a continuous and systematic process of collection, analysis, interpretation, and dissemination of descriptive information for monitoring health problems" (Rothman and Greenland 1998). Usually the surveillance systems are operated by public health officials to assist in disease prevention and to guide control.

All surveillance systems target a particular population, but this may range in size from a hospital or clinic to a community to a nation. The level of diagnosis and detection of disease in the community relates to the availability and accessibility of medical and laboratory services, ease of use, costs and time. In the case of *notifiable diseases*, health care providers are required to report specific infectious diseases in a particular population. The identification of "cases" may be based on various indicators of diseases, including symptoms or clinical descriptors, such as diarrhoea or fever (Haas et al. 1999), clinical diagnosis from testing of blood, faecal specimens, sputum, urine etc. for the pathogen, use of markers of infection (e.g. protein or genetic marker), and/or antibody (immunoglobulin) testing.

Public health surveillance (i.e., surveillance of health status and trends) contributes to the verification of drinking-water safety. Surveillance may assist in the identification of outbreaks or incidents of water-related disease or major threats, such as outbreaks or chemical contamination events. This relies on the provision of prompt and reliable information to public health authorities and strategies for protecting or advising members of the public or others at imminent threat of water-borne disease (Hunter et al. 2003).

However, there are a number of limitations in the use of surveillance for water-related illnesses. This includes the well-known "tip-of-the-iceberg" problem in detecting water-borne outbreaks, in which only a small fraction of true cases may be identified using the surveillance system (Hrudey and Hrudey 2004). The completeness of detection relies on the scope and efficiency of the surveillance system. Medical personnel may not notify the disease or may not identify that the disease originates from water. More commonly, laboratory isolates are used as the basis for disease detection, but this significantly underestimates the occurrence of disease because many people with illness do not seek medical help or get tested.

Wetlands as Settings for Human Health

There is a clear relationship between human health and wetlands.Horwitz and Finlayson (2011) identified nine ecosystem characteristics that can have consequences for human health:

- Contributions to hydration and safe water
- Contributors to nutrition
- Sites of exposure to pollution or toxicants
- Sites of exposure to infectious diseases
- Settings for mental health and psycho-social well-being
- Places from which people derive their livelihood
- Places that enrich people's lives, enable them to cope, and enable them to help others
- Sites of physical hazards
- Sites from which medicinal and other products can be derived.

As wetlands are potentially under threat from development concern needs to be given to the potential human health consequences of their development. For example, development of a wetland may alter biodiversity such that vector borne disease may increase, or removal of a wetland may remove a protein source for local communities.

It should be noted that in the event of human health being impacted by the alteration of a wetland ecosystem it may not necessarily be possible to restore the ecosystem to its previous state to reduce the human health risk (Weinstein 2005). In the restoration of an ecosystem the end point of the restoration may not be the original ecosystem state and different ecological processes may be at work (Suding et al. 2004) and therefore the ecosystem services provided may not be the same or at the same level.

Given the relationship between human health and wetland ecosystems Horwitz et al. (2012) highlighted the need to be vigilant for the emergence or re-emergence of wetland linked diseases, to act preventatively and proactively in relation to such diseases, and to develop scientifically-based responses, taking into account current good practices, where instances of such disease are identified. The unknown trajectory of degraded ecosystems or ecosystems being restored can have unexpected human health consequences. This requires a multidisciplinary approach by managers, incorporating public health and environment, as well as such dimensions as water supply, trade and development, but in particular an understanding of the linkages (and potential linkage) between human health and wetland ecosystems.

References

Aron JL, Patz J (2001) Ecosystem change and public health: a global perspective. Johns Hopkins University, Baltimore

Barling RD, Moore ID (1994) Role of buffer strips in management of waterway pollution-a review. Environ Manag 18:543–558

Barwick RS, Levy DA, Craun GF, Beach MJ, Calderon RL (2000) Surveillance for waterborne-disease outbreaks-United States, 1997–1998. Morbidity & mortality weekly report. CDC Surveill Summ 49:1–21

Corvalán C, Hales S, Mcmichael AJ, World Health Organization & Millennium Ecosystem Assessment (PROGRAM) (2005) Ecosystems and human well-being health synthesis: a report of the Millennium Ecosystem Assessment. Millennium ecosystem assessment. World Health Organization, Geneva

Cottingham KL, Chiavelli DA, Taylor RK (2003) Environmental microbe and human pathogen: the ecology and microbiology of Vibrio cholerae. Front Ecol 1:80–86

Crighton EJ, Elliot SJ, van der Meer J, Small I, Upshur R (2003) Impacts of an environmental disaster on psychosocial health and well-being in Karakalpakstan. Soc Sci Med 56 (3):551–567

Enhealth Council (2002) Environmental Health risk assessment: guidelines for assessing human health risks from environmental hazards. Environmental Health Systems Document 4. [Online]. Canberra: Department of Health and Ageing and enHealth Council, Commonwealth of Australia. http://enhealth.nphp.gov.au/council/pubs/ecpub.htm

European Commission Brussels (2003) Technical guidance document in support of commission directive 93/67/EEC on risk assessment for new notified substances and commission regulation (EC) No. 1488/94 on risk assessment for existing substances. European Commission

Fewtrell L, Bartram J (eds) (2001) Water quality guidelines, standards and health: assessment of risk and risk management for water-related infectious disease. IWA Publishing, London

Fossi MC, Marsili L (2003) Effects of endocrine disruptors in aquatic mammals. Pure Appl Chem 75:2235–2247

Giesy JP, Feyk LA, Jones PD, Kannan K, Sanderson T (2003) Review of the effects of endocrine-disrupting chemicals in birds. Pure Appl Chem 75:2287–2303

Gordis L (2009) Epidemiology. Elsevier/Saunders, Philadelphia

Gray SR, Becker NSC (2002) Contaminant flows in urban residential water systems. Urban Water 4:331–346

Guillette LJ, Iguchi T (2003) Contaminant-induced endocrine and reproductive alterations in reptiles. Pure Appl Chem 75:2275–2286

Guillette LJ, Edwards TM, Moore BC (2007) Alligators, contaminants and steroid hormones. Environ Sci 14:331–347

Haas CN, Rose JB, Gerba CP (1999) Quantitative microbial risk assessment. Wiley, New York

Hansen JS, Ongerth JE (1991) Effects of time and watershed characteristics on the concentration of Cryptosporidium oocysts in river water. Appl Environ Microbiol 57:2790–2795

Havelaar AH, Melse JM (2003) Quantifying public health risk in the WHO guidelines for drinking-water quality: a burden of disease approach. RIVM, Bilthoven

Hawkins PR, Swanson P, Warnecke M, Shanker SR, Nicholson C (2000) Understanding the fate of Cryptosporidium and Giardia in storage reservoirs: a legacy of Sydney's water contamination incident. J Water Supply Res Technol-Aqua 49:289–306

Hennekens CH, Buring JE, Mayrent SL (1987) Epidemiology in medicine. Little, Brown, Boston

Herwaldt BL, Craun GF, Stokes SL, Juranek DD (1991) Waterborne-disease outbreaks, 1989–1990. Morbidity & mortality weekly report. CDC Surveill Summ 40:1–21

Hodgson E (2004) A textbook of modern toxicology. Wiley, Hoboken

Horwitz P, Finlayson CM (2011) Wetlands as settings for human health:incorporating ecosystem services and health impact assessment into water resource management. Bioscience 61:678–688

Horwitz P, Finlayson M, Weinstein P (2012) Healthy wetlands, healthy people: a review of wetlands and human health interactions Geneva. Secretariat of the Ramsar Convention on Wetlands, The World Health Organisation, Switzerland

Hrudey SE, Hrudey EJ (2004) Safe drinking water: lessons from recent outbreaks in affluent nations. IWA Publishing, London

Hunter PR, Waite M, Ronchi E, Organisation for Economic Co-operation and Development (2003) Drinking water and infectious disease: establishing the links. CRC Press, Boca Raton

Jardine A, Speldewinde P, Lindsay MDA, Cook A, Johansen CA, Weinstein P (2008) Is there an association between dryland salinity and ross river virus disease in southwestern Australia? Ecohealth 5:58–68

Jobling S, Tyler CR (2003) Endocrine disruption in wild freshwater fish. Pure Appl Chem 75:2219–2234

Kolpin DW, Furlong ET, Meyer MT, Thurman EM, Zaugg SD, Barber LB, Buxton HT (2002) Pharmaceuticals, hormones, and other organic wastewater contaminants in US streams, 1999–2000: a national reconnaissance. Environ Sci Technol 36:1202–1211

Kramer MH, Herwaldt BL, Craun GF, Calderon RL, Juranek DD (1996) Surveillance for waterborne-disease outbreaks-United States, 1993–1994. Morbidity & mortality weekly report, CDC Surveill Summar 45:1–33

Lechevallier MW, Norton WD, Lee RG (1991) Occurrence of Giardia and Cryptosporidium Spp in Surface-Water Supplies. Appl Environ Microbiol 57:2610–2616

Leclerc H, Schwartzbrod L, Dei-Cas E (2002) Microbial agents associated with waterborne diseases. Crit Rev Microbiol 28:371–409

Lee SH, Levy DA, Craun GF, Beach MJ, Calderon RL (2002a) Surveillance for waterborne-disease outbreaks-United States, 1999–2000. Morbidity & mortality weekly report. Surveill Summar 51:1 47

Lee SH, Levy DA, Craun GF, Beach MJ, Calderon RL (2002b) Surveillance for waterborne-disease outbreaks-United States, 1999–2000. MMWR Surveill Summ 51:1–47

Levy DA, Bens MS, Craun GF, Calderon RL, Herwaldt BL (1998) Surveillance for waterborne-disease outbreaks-United States, 1995–1996. MMWR CDC Surveill Summ 47:1–34

Matthiessen P (2003) Historical perspective on endocrine disruption in wildlife. Pure Appl Chem 75:2197–2206

Mcfeters GA (1990) Drinking water microbiology: progress and recent developments. Springer, New York

Moore AC, Herwaldt BL, Craun GF, Calderon RL, Highsmith AK, Juranek DD (1993) Surveillance for waterborne disease outbreaks-United States, 1991–1992. MMWR CDC Surveill Summ 42:1–22

National Research Council (U.S.). Committee to Evaluate the Viability of Augmenting Potable Water Supplies with Reclaimed Water (1998) Issues in potable reuse: the viability of augmenting drinking water supplies with reclaimed water. National Academy, Washington, DC

National Water Quality, Management Strategy (2006) Australian guidelines for water recycling: managing health and environmental risks (Phase 1). Natural Resource Management Ministerial Council, Environment Protection and Heritage Council, Australian Health Ministers Conference

National Water Quality, Management Strategy (2008) Australian guidelines for water recycling: managing health and environmental risks (Phase 2), Augmentation of drinking water supplies. Natural Resource Management Ministerial Council, Environment Protection and Heritage Council, Australian Health Ministers Conference

Norris DE (2004) Mosquito-borne diseases as a consequence of land use change. EcoHealth 1:19–24

Oehlmann J, Schulte-Oehlmann U (2003) Endocrine disruption in invertebrates. Pure Appl Chem 75:2207–2218

O'Hara SL, Wiggs GFS, Mamedov B, Davidson G, Hubbard RB (2000) Exposure to airborne dust contaminated with pesticide in the Aral Sea region. The Lancet 355:627–628

Olson BH, Hurst CJ (1990) Microbiological aspects of drinking water. Session 134, Held Tuesday May 15 1990. Dallas, Tex.: Sound Solution

Organisation For Economic Co-Operation, and Development, World Health Organization (2003) Assessing microbial safety of drinking water: improving approaches and methods. OECD, Paris

Ostfeld RS, Keesing F (2000) The function of biodiversity in the ecology of vector-borne zoonotic diseases. Can J Zool 78:2061–2078

Ostfeld RS, Logiudice K (2003) Community disassembly, biodiversity loss, and the erosion of an ecosystem service. Ecology 84:1421–1427

Patz J (2007) Launch of the international association for ecology and health at its first biennial conference: message from the president elect. Ecohealth 4:6–9

Percival SL, Walker JT, Hunter PR (2000) Microbiological aspects of biofilms and drinking water. CRC Press, Boca Raton

Pruss A, Kay D, Fewtrell L, Bartram J (2002) Estimating the burden of disease from water, sanitation, and hygiene at a global level. Environ Health Perspect 110:537–542

Rodgers A, Ezzati M, Van der Hoorn S, Lopez AD, Lin RB, Murray CJ, Comparative Risk Assessment Collaborating G (2004) Distribution of major health risks: findings from the global burden of disease study. PLoS Med/Public Libr Sci 1:e27

Rothman KJ, Greenland S (1998) Modern epidemiology. Lippincott-Raven, Philadelphia

Rothman KJ, Greenland S, Lash TL (2008) Modern epidemiology. Wolters Kluwer Health/Lippincott Williams & Wilkins, Philadelphia

Speldewinde PC, Cook A, Davies P, Weinstein P (2009) A relationship between environmental degradation and mental health in rural Western Australia. Health Place 15:880–887

Speldewinde PC, Cook A, Davies P, Weinstein P (2011) The hidden health burden of environmental degradation: disease comorbidities and dryland salinity. Ecohealth 8:82–92

Suding KN, Gross KL, Houseman GR (2004) Alternative state and positive feedbacks in restoration ecology. Trend Ecol Evol 19:48–53

Szklo M, Nieto FJ (2007) Epidemiology: beyond the basics, Sudbury. Jones and Bartlett, Sudbury

Tebbutt THY (1998) Principles of water quality control. ButterWorth-Heinemann, Boston

Toze S (2006) Reuse of effluent water-benefits and risks. Agric Water Manag 80:147–159

Vörösmarty CJ, Mcintyre PB, Gessner MO, Dudgeon D, Prusevich A, Green P, Glidden S, Bunn SE, Liermann CR, Davies PM (2010) Global threats to human water security and river biodiversity. Nature 467:555–561

Weinstein P (2005) Human health is harmed by ecosystem degradation, but does intervention improve it? A research challenge from the Millenium Ecosystem Assessment. EcoHealth 2:228–230

WHO (2004a) Burden of disease and cost-effectiveness estimates. World Health Organisation

WHO (2004b) Guidelines for drinking water quality, 3rd edn. World Health Organization, Geneva

World Health Organization (1976) Surveillance of drinking-water quality. World Health Organization Geneva (obtainable from Q Corp)

World Health Organization (2005) Global assessment of the state-of-the-science of endocrine disrupters. WHO, Geneva

World Health Organization (2006a) Guidelines for drinking-water quality FIRST ADDENDUM TO THIRD EDITION, vol 1: Recommendations

World Health Organization (2006b) WHO guidelines for the safe use of wastewater, excreta and greywater. World Health Organization

World Health Organization (2006c) WHO guidelines for the safe use of wastewater, excreta and greywater/World Health Organization

Zetterstrom R (1998) Child health and environmental pollution in the Aral Sea region in Kazakhstan. Acta Paediatr Suppl 88:49–54

Wetlands and People's Well-being: Basic Needs, Food Security and Medicinal Properties

Anthony B. Cunningham

Abstract Demonstrating links between wetlands and health is a useful way of encouraging policymakers to act before the 'Water for Life' decade ends in 2015. This chapter describes the contributions wetlands make to people's well-being, such as food security through high water quality, protein or edible and medicinal plants. Earning cash income through trade in harvested wetland resources (such as fish, shell-fish or fibrous plants) can also have livelihood benefits. Although some wetlands can have negative effects on public health (such as bilharzia), public appreciation of positive wetland values and their links to public health can motivate local action and policy reform for wetland conservation and resource management. Lessons from innovative cases such as the RUPES programme in Indonesia can inspire new initiatives that put policies into practice for the benefit of local people.

Keywords Food security · Dietary diversity · Medicinal plants · Crop wild relatives · New natural products · Extremophiles

Introduction

It is well known that wetlands provide diverse and valuable goods and services for human well-being (Covich et al. 2004; "well-being" sensu Amatya Sen, see Berenger and Verdier-Chouchane 2007; Horwitz and Finlayson 2011). Despite this, wetland values are widely ignored and wetland destruction and degradation are widespread. The reason for our behaving this way is widely recognized. The direct and indirect values of wetlands are not taken into account, property rights are weak and wetlands can be affected by degradation elsewhere in the watershed (Dudgeon et al. 2006; Turner and Jones 1991). Although wetlands represent a capital asset providing important ecosystem services, these assets are generally not reflected in conventional economic indicators. Instead, wetland goods and services are often considered a "subsidy from nature", just as is the case with forests (Wunder 2007). Integrating economic (Barbier et al. 1997) and socio-ecological approaches (Berkes

A. B. Cunningham (✉)
School of Plant Biology, University of Western Australia, 6009 Nedlands, Australia
e-mail: tonyc05@bigpond.net.au

© Springer Science+Business Media Dordrecht 2015
C. M. Finlayson et al. (eds.), *Wetlands and Human Health*, Wetlands: Ecology, Conservation and Management 5, DOI 10.1007/978-94-017-9609-5_3

and Folke 1998) in policy and practice for continued provision of those wetland goods and services is therefore essential. Valuation estimates can be controversial (Balmford et al. 2002). In addition, non-market values, such as cultural or religious values of ecosystems are extremely difficult to determine (Adamowicz et al. 1998). Values can also be location specific, such as the high value placed on *Juncus kraussii* culms in southern Africa (Cunningham and Terry 2006) compared to Australia, where this species also occurs.

Despite controversies and gaps, valuation studies commonly indicate the great economic importance of wetlands. What is needed is more effective communication by scientists to policymakers on the importance of wetlands to people's well-being. Links between direct and indirect drivers of wetland change and opportunities provide guidance on how these link to human well-being (Fig. 1).

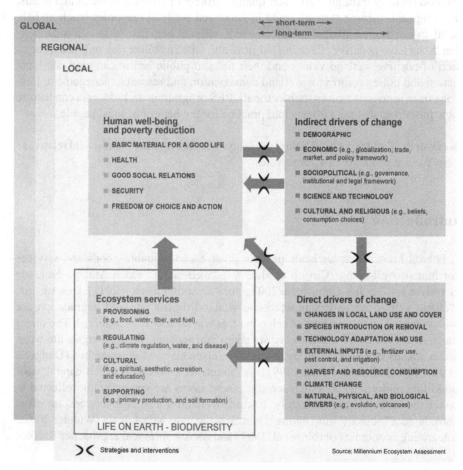

Fig. 1 Links between ecosystem services, drivers of change, human well-being and poverty reduction. (from Millennium Ecosystem Assessment 2005)

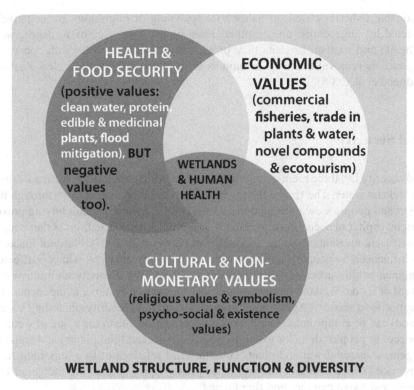

HEALTH &
FOOD SECURITY
(positive values:
clean water, protein,
edible & medicinal
plants, flood
mitigation), **BUT**
negative
values
too).

WETLANDS
& HUMAN
HEALTH

**ECONOMIC
VALUES**
(commercial
fisheries, trade in
plants & water,
novel compounds
& ecotourism)

CULTURAL & NON-
MONETARY VALUES
(religious values & symbolism,
psycho-social & existence
values)

WETLAND STRUCTURE, FUNCTION & DIVERSITY

Fig. 2 Links between wetlands and our health and well-being

This chapter describes the contributions wetlands make to people's well-being (Fig. 2). These range from wetland ecosystem services (such as clean water), economic productivity and poverty alleviation (wetlands and fisheries) to food security (such as the genetic diversity of wild relatives of rice (*Oryza*), one the world's major crops, or new natural products from wetland associated fungi, bacteria, animals and medicinal plants). In addition, some wetlands have "insurance" value, reducing our vulnerability to extreme events such as floods while others, such as peatlands, play an important role in carbon sequestration.

Although the different contributions wetlands make to people's well-being are discussed separately below, there are linkages within and between each one. Food security, for example, links to water quality, household income, plant genetic resources and fisheries management. These linkages mean that in many cases, trade-offs between wetland conservation and development need to be carefully assessed and in some instances, compromises reached. Strategies in Bangladesh to reach trade-offs between floodplain conversion for rice production or wetland maintenance for survival and production of floodplain fish populations when both are important in the peoples diet are a good example (Shankar et al. 2004). Similarly, important trade-offs exist between the benefits of development projects such as large dams on one hand, and infectious disease risk (such as schistosomiasis) or sustainable fisheries based on fish whose spawning or migrations are negatively impacted by large dams, on the other. Even recreational uses of wetlands, with the health and tourism benefits they provide may need trade-offs with conservation, such as removal of exotic salmonids from rivers in Chiles and New Zealand (Dudgeon et al. 2006).

Food Security

Food security has three main components, each of which has links to wetland values for food and water. The first is the most obvious, the food availability (through the market and people's own production). The second is having enough buying power or social capital to access food with cash or through barter. The third is that people get sufficient nutrients from the food they eat (Boko et al. 2007). Nutrient intake is also influenced by people's ability to digest and absorb nutrients, which is affected by human health, access to safe drinking water and the diversity and nutritional content of foods. Wetland degradation or loss impacts on all three components that comprise food security. Agricultural production and food security (including access to food) can be compromised with wetland degradation due to the value of wetland resources to peoples diet (for example from fisheries and food plants) and some of the genetic material wetland plants contain. Wild relatives of two important food crops, rice (*Oryza* species) and some cowpeas (*Vigna*) species are indigenous to African and Asian marshes and floodplains.

Starchy staple diets (rice, cassava, maize) are frequently deficient in nicotinic acid, vitamin C, calcium and riboflavin and protein (Cunningham and Shackleton 2004). Harvested wild foods are known to be a valuable source of these nutrients deficient in starchy staple diets, particularly protein from edible fish and shellfish, nicotinic acid from wild edible greens, vitamin C from wild fruits. Although many edible wild plants are harvested from forests or woodlands rather than wetlands, a wide diversity of wetland plants provide supplementary food sources. Popular food species are also traded. Examples are water-cress (*Rorippa nasturtium-aquaticum*) in Europe, *Mauritia* fruits and *Euterpe* palm hearts from South America floodplain forests, lotus (*Nelumbo nucifera*) seeds and water-chestnut (*Eleocharis*) tubers in Asia and wild rice (*Zizania aquatica*) and cranberries (*Vaccinium oxycoccos*) collected for food and trade by native Americans in the USA.

Fish are particularly important to people's diet and health in developing countries where they often form the main source of animal protein (Fig. 3). In recent years, the production of fish from inland waters has been dominated by aquaculture, especially growing of carp in China for domestic purposes, with salmon, tilapia, and perch mainly for export (Kura et al. 2004). Inland fisheries are particularly important in developing countries with a large proportion of the recorded catch, with the actual catch possibly several times higher than recorded (Kura et al. 2004). Most of the increase in freshwater fish consumption occurred in Asia and Africa. Historical date for commercial fisheries shows large declines in the twentieth century due to habitat degradation, invasive species, and overharvesting (Revenga et al. 2000).

Fig. 3 Small-scale fish harvesting using traditional methods, often by women and children provides much needed protein—in this case along the middle-Zambezi river, Zambia. (Photo: A B Cunningham)

The Mekong river, for example, sustains one of the world's largest freshwater fisheries, with annual yields of 1 million tonnes of fish, most of which are harvested by small-scale artisanal fisheries (Valbo-Jorgenson and Poulsen 2001). In Cambodia, for example, people get about 60–80% of their total animal protein from the fishery in Tonle Sap and associated floodplains (MEA 2005). Floodplain fisheries are often very productive, although fish production is highly variable due to seasonal floods and longer-term climatic trends (Jul-Larsen et al. 2003) that threaten fisheries such as those around Lake Chad.

Water Supplies

Fresh water is a basic need for human health. This is widely recognized in national legislation of many countries. This provides an important opportunity for linking local action and public health to wetlands conservation. In the USA, for example, the Clean Water Act has become an important tool enabling Native Americans to leverage wetland conservation and restoration at a catchment level (USEPA 2000). On average, people need 20–50 litres of clean water per person per day for drinking, cooking and personal hygiene, yet over 1 billion people lack access to safe water supplies and 2.6 billion people lack adequate sanitation (MEA 2005). Wetland vegetation plays an important role in improving water quality through extraction of pollutants and pathogens including nitrates, coliform bacteria and faecal streptococci (Ghermandi et al. 2007). In fact this role is so useful that artificial wetlands have been purposely created for this purpose in France for over 20 years (Molle et al. 2005), with the design principles behind riparian wetland construction being developed for stream restoration (D'Arcy et al. 2007). Poor quality water contributes to a range of health problems such as diarrhoea, internal parasites and trachoma. Bad health due to lack of access to safe drinking water and poor sanitation affects the poorest sector of society, with follow-on affects for food security.

Shelter

Building styles and the materials used reflect cultural diversity and preferences for particular species, as well as what is available from vegetation change. In many developing countries, where house construction reflects need rather than restrictive building codes of the developed world, locally harvested plants are the main source of low-cost housing. Although hardwoods from upland forests and woodlands are preferred for support poles, wetlands are a favoured source of thatching material and reeds (Cunningham 1985). In Africa, floodplain grasses and Cyperaceae are commonly used for thatching traditional houses. In southern Africa, the common reed (*Phragmites australis*) was used for wall construction of up to 90 % of homes (Cunningham 1985) and in Europe, the same species is used for expensive thatch.

Subsistence Income

For rural people wanting to enter the cash economy, harvesting wild resources (salt, fish, shell-fish, useful plants) is an important option, as local knowledge and skills can be used to harvest products for trade without an initial investment of cash. Complex trade networks commonly characterize this hidden economy. As mentioned earlier, buying power can also help with food security. In many developing countries, these resources also provide a "green social security", as unlike Europe, Australia and North America, there are no government social security payments in times of need. Although income from harvest and trade is small by western standards, its values to households should not be underestimated. Trade in fresh, dried or smoked fish is widespread through Asia, Africa (Abbott et al. 2007) and Latin America. So too, is trade in basketry, including fish traps.

While bamboo and rattans from upland forests and agroforestry systems are a common source of basketry fibres in Asia, plants from wetland and high water-table palm savannas dominate African basketry fibres. The development of commercial craft enterprises since the 1970s has brought much needed income for producers and their families, where it is often used for school costs (Cunningham and Terry 2006). Most southern African basket makers are women from low-income families, living in remote rural areas, and are subsistence farmers who own few (if any) cattle, and have little or no education. For most weavers, cash income is obtained through the sale of home-brewed beer, grain, bread, or thatching grass; casual labour or employment on public works; old age pensions; or money sent by family members who are migrant workers. For many, the only consistent source of cash income is through the production and sale of handicrafts, especially baskets. As a male basket maker in Khwai, Botswana, stated, '*My baskets are my cattle*'. Cultural values can also drive commercial harvest. Each year, several thousand Zulu women harvest mat rush (*Juncus kraussii*) from 20 hectare area of coastal salt-marsh at St Lucia estuary, a Ramsar site in South Africa. These culms are then resold or made into sleeping mats prized for their cultural significance at weddings or crafts for export, with intensive use causing concern about sustainability (Heinsohn and Cunningham 1991).

Traditional Medicines and New Natural Products

Although traditional medicines are dominated by flowering plant use (most of them not from wetlands), it is wetland associated animals (such as leeches and frogs), fungi, bacteria and extremophile lower plants (algae) (e.g. Goss 2000) rather than flowering plants that provide the most productive sources of new natural products. In terms of people's health, both sectors need to be considered. In some cases, there are close links between the new and old uses of organisms, sometimes from different wetlands on different continents. The medicinal leech (*Hirudo medicinalis*) from European freshwater wetlands provides a good example. Traditionally used for bleeding patients in medieval Europe, leeches are now the source of hirudin, the first major new anticoagulants brought into health care since heparin was discovered in the early 1900s (Moreal et al. 1996). The link between old and new doesn't end there. To produce sufficient quantities of heparin for therapeutic use requires recombinant technology. This is done using bacteria, eukaryotes and yeasts to produce recombinant forms of hirudin (r-hirudin) (Sohn et al. 2001). Taq polymerase, widely used in polymerase chain reaction (PCR) technology, including DNA sequencing into the genetic material of another organism, is from DNA polymerase of *Thermus aquaticus*, a bacterial "extremophile" which occurs in the geysers of Yellowstone National Park, where its ability to survive extreme heat enables its DNA polymerase to survive the successive heating cycles of PCR. Aside from the direct health and economic values of hirundin is the value of the technology developed from *Thermus aquaticus*. Not only did this win its inventor, Karry Mullis, the Nobel Prize in 1993, but in 1991, the Swiss pharmaceutical company Hoffman-Laroche bought the exclusive world rights to the PCR process for US$ 300 million from Cetus Corporation, for whom Karry Mullis worked at the time (Doremus 1999). In 2005, worldwide sales of PCR enzymes were reported to be in the range of US$ 50–100 million (Lohan and Johnson 2005) and may be more today, given growth in the biotechnology field.

These examples illustrate several points relevant to the confluence between wetlands, the Ramsar Convention on Wetlands, natural products and human health. Firstly, the medicinal qualities of leeches are a good example of the continued value of traditional knowledge to health care today. Secondly, new technologies, such as rapid throughput screening (White 2000) and PCR are changing the face of new natural product development. Thirdly, links between wetland biodiversity and human health need to focus less on the obvious (such as birds, large mammals or plants), than on the "hidden biodiversity" (such as fungi and bacteria). Fourthly, the case of biodiversity prospecting for *Thermus aquaticus* illustrates how controversial this can be, with important policy implications and links to the Convention on

Biodiversity (CBD). Finally, the most likely places for promising leads are wetland species from environments such as hot springs, alpine wetlands, particularly in high diversity montane systems, including the Andes or Himalaya, desert salt-pans, soda lakes, highly alkaline or acid streams and high diversity tropical rivers. Many of these are not listed as internationally important under the Ramsar Convention, although there are exceptions, such as the hot springs and soda lakes of East Africa's Rift Valley (Lake Bogoria and Lake Elementeita). Given that few Ramsar listed wetlands are located on mountains or in deserts, compared to lowlands and along the coast, it may be worth considering the addition of wetlands from other environments to support several goals, including biodiversity conservation.

Traditional Medicines

Worldwide, the skewed distribution of medical doctors is a weakness in public healthcare. Typically, high numbers of medical doctors practice in large cities of developed countries and low numbers in rural areas of developing countries (Wibulpolprasert and Pengpaibon 2003). As a result, traditional medicines continue to serve as the main form of health care for an estimated 80 % of people in developing countries (WHO 2002). Across the world, diverse local health care systems have developed over hundreds, or thousands of years through complex and dynamic interactions between people and their environment, commonly used to treat parasitic diseases, diarrhoea, and for oral hygiene. Use of medicinal plants is also widespread in developed countries. In Australia, for example, 48 % of people use complementary and alternative medicine (CAM) and 42 % of the population in the United States reportedly use CAM (Eisenberg et al. 1998), with use levels increasing significantly over the recent past (Schippmann et al. 2003) (Fig. 4).

Worldwide it is estimated that of 422,000 flowering plants, 12.5 % (52,000) are used medicinally with 8 % (4160 species) of these threatened (Schippmann et al. 2003). At a global scale, export of medicinal and aromatic plants to China, India and Germany is huge, with China the largest exporter mainly to Hong Kong, (140,500 t) as well as being the world's major importer (80,550 t) (Lange 1998). Medicinal properties of plants are commonly concentrated in particular plant families, reflect their evolutionary history and ecological adaptations, such as chemical defenses against herbivores, fungi or pathogens. Although seeds from common wetland plants such as cattail (*Typha*), common reed (*Phragmites*) and lotus (*Nelumbo nucifera*) are widely used in traditional medical systems, wetlands dominated by monocotyledons (Cyperaceae, Juncaceae, Typhaceae, Poaceae) are a far less important source of medicinal plants than flooded forests, swamp forests and mountain wetlands and seepage areas. Many of China and India's most important medicinal plants, for example, are from montane bogs, seepage areas and alpine pastures of the Himalaya rather than the coastal systems better represented by Ramsar listed wetlands. Nepal, for example exports between 7000 and 27,000 t of medicinal plants a year, most of them to India, worth between US$ 7–30 million/year (Olsen 2005). Many of these

Fig. 4 A traditional medicines market in Xi'an, China, where, many plants and animals, including species from wetlands, continue to contribute to health care. (Photo: A B Cunningham)

are montane medicinal plants, including threatened species, the Ranunculaceae (*Aconitum*), Papaveraceae (*Meconopsis*), Scrophulariaceae (*Picrorhiza*) and Valerianceae (*Nardostachys*). Exceptions to the limited number of medicinal plants in lowland systems are the flooded forests and swamp forests of the African, Asian and South American lowland tropics, which contain a high diversity of medicinal trees and shrubs in the Apocynaceae (*Rauvolfia, Tabernaemontana*), Clusiaceae (*Clusia, Garcinia*), Rubiaceae (*Genipa*) and Euphorbiaceae (*Phyllanthus*).

In Asia, particularly China, India, Pakistan and Vietnam, government support for the development and modernization of traditional medical systems is likely to increase harvest levels from wild stocks. In India, where the Ayurvedic industry is worth an estimated US$ 1 billion per year, 7500 factories produce thousands of Ayurvedic and Unani formulae (Bode 2006). In China, clinical trials for TCM preparations are now frequent (Qiong et al. 2005) and the plan is to establish a series of standards for modern TCM products and a competitive modern TCM industry through new technology and standardization. In Africa and South America, production is less formalised and branding less sophisticated, yet the scale of the trade is deceptively large. In South Africa, for example, 1.5 million informal sector traders sell about 50,000 tonnes of medicinal plants annually in a region with an estimated 450,000 traditional healers (Mander 2004). In common with China, India and Nepal, relatively few medicinal species in African and Madagascar trade are from wetlands. Notable exceptions are a massive trade in endemic *Drosera madagascariensis* (Drosearaceae) from Madagascar to Europe (Paper et al. 2005), and in southern Africa where several species from montane marshes and seepages, such as *Allepidea amatymbica* (Apiaceae) are used for coughs and *Gunnera perpensa* (Gunneracae) which is used in herbal preparations prior to childbirth. Many wild species supplying medicinal plant markets are declining in their availability, with important implications for primary health care (Cunningham 1993).

New Natural Products

New natural products discovery have been radically changed due to the availability of molecular biology, PCR technology (thanks for *Thermus aquaticus* and innovative research) and genomic sciences (Drews 2000). In many ways, the biotechnology industry has become a major tool of the industry. Although the focus of this chapter is human health, new natural products have a wide range of other applications, from agriculture to cosmetics, including some with direct links to habitat conservation. The fungal infection, *Phytophthora*, for example, poses the major conservation threat to south-western Australia's unique flora. One of the active ingredients used to treat *Phytophthora*, known as oocydin A, which has application in agriculture and forestry and conservation restoration was developed from *Rhyncholacis penicillata* (Podostemaceae), a plant from rivers in South-west Venezuela associated with an endophytes *Serratia marcescens* which produced oocydin A, a novel anti-oomycetous compound (Strobel et al. 1999).

New antibiotics are a good example of health links to new natural products, with 5000–10,000 new antibiotics discovered from bacteria and fungi since the 1950s and 1960s when well known drugs such as tetracycline were discovered (Challis and Hopwood 2003). The bulk of these have come from *Streptomyces* species, which are saprophytes found in soil, marine sediments and plant tissues. Endophytic microorganisms, which are commonly found on plants, including many wetland species produce a diverse range of compounds with potential use in medicine, agriculture and industry, including new antibiotics, anti-mycotics, immuno-suppressants and anti-cancer compounds (Strobel and Daisy 2003). The most promising wetlands to search for endophytes with commercial potential are high diversity systems of tropical lowlands, montane and boreal systems rather than mono-dominant wetlands. Recent studies in Canadian wetlands are a good example of this (Kuhajek et al. 2003). Implementation of the Convention on Biological Diversity's policies on access and benefit sharing are important to recognize as the search for new natural products continues. These have been outlined recently in the Nagoya Protocol (Secretariat of the Convention on Biological Diversity 2012).

In addition to *Thermus aquaticus*, as the best known extremophile, there is great interest in other extremophiles. Wetland examples are the green algae *Dunaliella acidophila*, which survives at pH 0 and *Gloeochrysis* which lives on stones in acidic (pH 2) streams running out of active volcanoes in Patagonia, Argentina (Goss 2000; Baffico et al. 2004). The industrial applications of natural products from these extremeophiles include waste treatment, liposomes for drug delivery and cosmetics, and the food industry. This can have both positive outcomes (such as waste treatment) and negative outcomes for wetlands and human health (such as their use in protein-degrading additives in detergents, made possible due to their ability to with stand high temperatures).

Conclusion

In many parts of the world, indigenous and local peoples have existed in harmony with wetlands for centuries. In urban-industrial societies this is often not the case, resulting in adverse impacts not only on both wetlands and people's well-being. Understanding the links between ecosystem services and human health, as detailed by Horwitz and Finlayson (2011) is a crucial entry point for improvements in policy and practice for wetland conservation and restoration. Maintaining or restoring wetland goods and services cannot be achieved by working in isolation, but has to be achieved on the basis of entire watersheds. Achieving this is complex, even on a national scale, but can be done. A recent process that could be followed is the adaptive co-management system developed for a Ramsar listed wetland in the lower Helgeå River catchment, Sweden (Olsson et al. 2004). As wetland goods and services become scarcer, interest in the idea of paying others, such as communities on forested land, to provide ecosystem services on a sustained basis, is also growing (Katoomba Group 2007; Wunder 2007; Horwitz and Finlayson 2011). Worldwide, payment for ecosystem services (PES) is at an early stage, so as would be expected, fewer projects were identified in this recent inventory where money had exchanged hands. In Asia, PES schemes relevant to Ramsar are also growing. The first example is watershed management projects under RUPES (Rewarding the Upland Poor for Environmental Service) in Philippines, Nepal, Indonesia (Swallow et al. 2005). The second case is at a much larger scale, costing 3.65 billion yuan (c. US$ 2.4 billion) between 1999 and 2001. Planned to reduce soil erosion from steep slopes in catchments, the "Grain for Green" programme in China has involved nearly 15 million ha of cropland and 40–60 million rural households (by 2010) (Ushida et al. 2005).

Wetland restoration using ecological engineering is also being implemented in many parts of the world (Alexander and McInnes 2012). Since the wake-up call from hurricane Katrina, good science is also being applied to re-establish ecosystem services and reconnect the Mississippi river to the deltaic plain (Covich et al. 2004). The 'Water for Life' decade ends in 2015. Now is the time to effectively communicate links between wetlands and health to get policymakers to act.

References

Abbott JG, Campbell LM, Hay CJ, Næsje TF, Purvis J (2007) Market-resource links and fish vendor livelihoods in the Upper Zambezi River floodplains. Hum Ecol 35:559–574
Adamowicz W, Beckley T, Hatton Macdonald D, Just L, Luckert M, Murray E, Phillips W (1998) In search of forest resource values of indigenous peoples: are non-market valuation techniques applicable? Soc Nat Resour 11:51–66

Alexander S, McInnes R (2012) The benefits of wetland restoration. Ramsar Scientific and Technical Briefing Note no. 4. Gland. Ramsar Convention Secretariat, Switzerland

Baffico GD, Diaz MM, Wenzel MT, Koschorreck M, Schimmele M, Neu TR, Pedrozo F (2004) Community structure and photosynthetic activity of epilithon from a highly acidic (pH=2) mountain stream in Patagonia, Argentina. Extremophiles 8:463–473

Balmford A, Bruner A, Cooper P, Constanza R, Farber S, Green RE, Jenkins M, Jefferiss P, Jessamy V, Madden J, Munro K, Myers N, Naeem S, Paavola J, Rayment M, Rosendo S, Roughgarden J, Trumoer K, Turner RK (2002) Economic reasons for conserving wild nature. Science 297:950–953

Barbier E, Acreman M, Knowler D (1997) Economic valuation of wetlands—a guide for policy makers. Ramsar Convention Bureau/IUCN, Gand, Switzerland. http://www.biodiversityeconomics.org/valuation/topics-02-00.htm

Berenger V, Verdier-Chouchane A (2007) Multidimensional measures of well-being: standard of living and quality of life across countries. World Dev 35:1259–1276

Berkes F, Folke C (eds) (1998) Linking social and ecological systems: management practices and social mechanisms for building resilience. Cambridge University Press, Cambridge

Bode M (2006) Taking traditional knowledge to the market: the commoditization of Indian medicine. Anthropol Med 13:225–236

Boko M, Niang I, Nyong A, Vogel C (2007) Chapter 9: Africa. IPCC WGII Fourth Assessment Report

Challis GL, Hopwood DA (2003) Synergy and contingency as driving forces for the evolution of multiple secondary metabolite production by Streptomyces species. Proc Nat Acad Sci U S A 100:14555–14561

Covich AP, Ewel KC, Hall RO, Giller PE, Goedkoop W, Merritt DM (2004) Ecosystem services provided by freshwater benthos. In: Wall DH (ed) Sustaining biodiversity and ecosystem services in soil and sediments. Island Press, Washington, DC, pp 45–72

Cunningham AB (1985) The resource value of indigenous plants to rural people in a low agricultural potential area. Faculty of Science, University of Cape Town

Cunningham AB (1993) African medicinal plants: setting priorities at the interface between conservation and primary health care. People and Plants Working Paper vol 1. UNESCO, Paris, pp 1–50

Cunningham AB, Shackleton CM (2004) Use of fruits and seeds from indigenous and naturalized plant species. Chapter 20. In: Eeeley H, Shackleton C, Lawes M (eds) Use and value of indigenous forests and woodlands in South Africa. University of Natal Press, Pietermartizburg

Cunningham AB, Terry ME (2006) African basketry: grassroots art from southern Africa. Fernwood Press, Cape Town

D'Arcy BJ, MacLean N, Heal KV, Kay D (2007) Riparian wetlands for enhancing the self-purification capacity of streams. Water Sci Technol 56:49–57

Doremus H (1999) Nature, knowledge and profit: the Yellowstone's bioprospecting controversy and the core purposes of America's national parks. Ecol Law Q 26:402–405

Drews J (2000) Drug discovery: a historical perspective. Science 287:1960–1964

Dudgeon D, Arthington H, Gessner MO, Kawabata Z-I, Knowler DJ, Leveque C, Naiman RJ, Prieur-Richard A-H, Soto D, Stiassny MLJ, Sullivan CA (2006) Freshwater biodiversity: importance, threats, status and conservation challenges. Biol Rev 81:163–182

Eisenberg DM, Davis RB, Ettner SL et al (1998) Trends in alternative medicine use in the United States, 1990–1997: results of a follow-up national survey. JAMA 280:1569–1575

Ghermandi A, Bixio D, Traverso P, Cersosim I, Thoeye C (2007) The removal of pathogens in surface-flow constructed wetlands and its implications for water reuse. Water Sci Technol 56(3):207–216

Goss W (2000) Ecophysiology of algae living in highly acidic environments. Hydrobiologia 433:31–37

Heinsohn D, Cunningham AB (1991) Utilization and potential for cultivation of the salt-marsh rush *Juncus kraussii*. S A J Bot 57(1):1–5

Horwitz P, Finlayson CM (2011) Wetlands as settings for human health: incorporating ecosystem services and health impact assessment into water resource management. Bioscience 61:678–688

Jul-Larsen E, Kolding J, Overå R, Nielsen J, Zwieten P (2003) Management, co-management or no management? Major dilemmas in Southern Africa freshwater fisheries. FAO Fisheries Technical Paper 426/1. Food and Agriculture Organization, Rome

Katoomba Group (2007) Current 'State of Play' of carbon, water, and biodiversity markets. The full text of the inventories can be found at: http://www.katoombagroup.org/africa/pes.htm

Kuhajek JM, Clark AM, Slattery M (2003) Ecological patterns in the antifungal activity of root extracts from rocky mountain wetland plants. Pharm Biol 41:522–530

Kura Y, Revenga C, Hoshino E, Mock G (2004) Fishing for answers. World Resources Institute, Washington, DC, p 138

Lange D (1998) Europe's medicinal and aromatic plants. Their use, trade and conservation. TRAFFIC International, Cambridge, 77 p

Lohan D, Johnston S (2005) Bioprospecting in Antarctica. United Nations University Institute of Advanced Studies (UNU-IAS), Yokohama

Mander M (2004) Phytomedicines industry in southern Africa. In: Diederichs N (ed) Commercialising medicinal plants: a Southern African guide. African Sun Media, Pretoria

MEA (Millennium Ecosystem Assessment) (2005) Ecosystems and human well-being: Wetlands and water Synthesis. World Resources Institute, Washington, DC

Molle P, Lienard A, Boutin C, Merlin G, Iwema A (2005) How to treat raw sewage with constructed wetlands: an overview of the French systems. Water Sci Technol 51:11–21

Moreal M, Costa J, Salva P (1996) Pharmacological properties of hirudin and its derivatives. Potential clinical advantages over heparin. Drugs Aging 8:171–182

Olsen CS (2005) Valuation of commercial central Himalayan medicinal plants. Ambio 34:607–610

Olsson P, Folke C, Hahn T (2004) Social-ecological transformation for ecosystem management: the development of adaptive co-management of a wetland landscape in southern Sweden. Ecol Soc 9(4):2

Paper DH, Karall E, Kremser M, Krenn L (2005) Comparison of the anti-inflammatory effects of *Drosera rotundifolia* and *Drosera madagascariensis* in the HET-CAM assay. Phytother Res 19(4):323–326

Qiong W, Yiping W, Jinlin Y, Tao G, Zhen G, Pengcheng Z (2005) Chinese medicinal herbs for acute pancreatitis. Cochrane Database Syst Rev 1: CD 003631 1–14

Revenga C, Brunner J, Henninger N, Kassem K, Payne R (2000) Pilot analysis of global ecosystems: freshwater systems. World Resources Institute, Washington, DC, p 83

Schippmann U, Leaman DJ, Cunningham AB (2003) Impact of cultivation and gathering of medicinal plants on biodiversity: global trends and issues. Case study no. 7. In: Biodiversity and the ecosystem approach in agriculture, forestry and fisheries. Proceedings: Satellite event on the occasion of the Ninth Regular Session of the Commission on Genetic Resources for Food and Agriculture, Rome 12–13 October 2002. FAO, Rome. ISBN 92-5-104917-3. http://www.fao.org/DOCREP/005/Y4586E/y4586e08.htm#P1_0

Secretariat of the Convention on Biological Diversity (2012) Nagoya protocol on access to genetic resources and the fair and equitable sharing of benefits arising from their utilization to the convention on biological diversity: text and annex

Shankar B, Halls A, Barr J (2004) Rice versus fish revisited: on the integrated management of floodplain resources in Bangladesh. Nat Resour Forum 28:91–101

Sohn JH, Kang HA, Rao KJ, Kim CH, Choi ES, Chung BH, Rhee SK (2001) Current status of the anticoagulant hirudin: its biotechnological production and clinical practice. Appl Microbiol Biotechnol 57:606–613

Strobel G, Daisy B (2003) Bioprospecting for microbial endophytes and their natural products. Microbiol Mol Biol Rev 67:491–502

Strobel GA, Li JY, Sugawara F, Koshino H, Harper J, Hess WM (1999) Oocydin A, a chlorinated macrocyclic lactone with potent anti-oomycete activity from *Serratia marcescens*. Microbiology 145:3557–3564

Swallow B, Meinzen-Dick R, van Noordwijk M (2005) Localizing demand and supply of environmental services: interactions with property rights, collective action and the welfare of the poor. CAPRi Working Paper # 42., Washington, DC. www.worldagroforestrycentre.org/sea/Publications/files/workingpaper/WP0054-05.PDF

Turner RK, Jones T (eds) (1991) Wetlands, market and intervention failures. Earthscan, London

USEPA (United States Environmental Protection Agency) (2000) Tribal wetland program highlights. EPA 843-R-99-002. Office of Water, Office of Wetlands, Oceans and Watersheds, Washington, DC 20460

Ushida E, Xu J, Roselle S (2005) Grain for green: cost-effectiveness and sustainability of China's conservation set-aside program. Land Econ 81:247–264

Valbo-Jorgensen J, Poulsen AJ (2001) Using local knowledge as a research tool in the study of river fish biology: experiences from the Mekong. Environ Dev Sustain 2:253–276

White RE (2000) High-Throughput screening in drug metabolism and pharmacokinetic support of drug discovery. Ann Rev Pharmacol Toxicol 40:133–157

WHO (2002) Traditional medicine strategy 2002–2005. World Health Organisation, Geneva, Switzerland. www.who.int/medicines/library/trm/trm_strat_eng.pdf

Wibulpolprasert S, Pengpaibon P (2003) Integrated strategies to tackle the inequitable distribution of doctors in Thailand: four decades of experience. Hum Resour Health 1:12 [abstract only]

Wunder S (2007) The efficiency of payments for environmental services in tropical conservation. Conserv Biol 21(1):48–58

Wetlands as Sites of Exposure to Water-Borne Infectious Diseases

Bonnie T. Derne, Philip Weinstein and Colleen L. Lau

Abstract Wetlands provide many essential and important ecosystem services to humans, resulting in our considerable reliance on and exposure to various wetland environments. A subset of microorganisms and invertebrates commonly found in wetlands can cause diseases in humans, some of which are responsible for significant disease burden globally. As past disease outbreaks and emergence arising from wetlands have shown, the combination of predicted intensification of extreme weather events, human needs for land and natural resources, and biodiversity loss in the future are likely to drive the transmission of infectious diseases, resulting in an increasing burden of water-borne diseases, particularly where sanitation infrastructure is poor. The importance of preventing contamination, providing adequate sanitation and preserving or restoring healthy, service-providing ecosystems as strategies for risk mitigation are therefore underlined. A greater understanding of the complexities of wetland ecosystems and the interlinked environmental, microbiological and human factors that lead to infection risk should therefore be an objective of future research.

Keywords Infectious disease · Disease emergence · Disease outbreaks · Environmental health · Public health · Water-borne diseases · Ecosystems · Cholera · leptospirosis · Schistosomiasis · Giardia · Cryptosporidium · Biodiversity

Human Exposure to Wetlands

Wetlands encompass a diverse range of water-based environments, upon which humans are heavily reliant for the provision of resources, and the site of many economic and cultural activities. For these reasons, most communities around the world are located in or near wetlands.

C. L. Lau (✉) · B. T. Derne
Queensland Children's Medical Research Institute, Brisbane, Australia, WHO Collaborating Centre for Children's Health and Environment, The University of Queensland, Brisbane, Australia, Royal Children's Hospital, Herston, QLD 4029, Australia
e-mail: colleen.lau@uq.edu.au

P. Weinstein
School of Biological Sciences, The University of Adelaide, Adelaide 5005, Australia
e-mail: philip.weinstein@adelaide.edu.au

© Springer Science+Business Media Dordrecht 2015   45
C. M. Finlayson et al. (eds.), *Wetlands and Human Health,* Wetlands: Ecology, Conservation and Management 5, DOI 10.1007/978-94-017-9609-5_4

In this chapter, the term 'wetlands' includes both inland water systems such as lakes, rivers, streams, swamps, marshes, cultivation fields, and also shallow coastal environments such as estuaries, and mangrove systems. Wetlands are distributed throughout the world, though water-borne infectious diseases associated with these environments have the largest impact in low-income countries, as discussed below.

Wetlands provide many essential or important ecosystem services to humans (Corvalan et al. 2005; Horwitz et al. 2012). Chief amongst these services is the provision of water for drinking and other purposes, food, livelihood, and building materials. Wetlands also provide regulatory services, such as buffering from storms or floods, filtering of nutrients and pollutants, carbon sequestration and the regulation of disease causing pathogens. The cultural identity of many people and communities is shaped by wetlands, which are also settings for recreational, cultural, and spiritual activities important for psychosocial well-being. Due to this strong, multifaceted reliance on wetlands, humans have significant exposure to them. Exposure can be direct (e.g. spending time in the wetlands) (Fig. 1), or indirect if water and food originate from wetlands. Developing world and rural populations are likely to have more direct exposure to and a greater degree of reliance on wetlands, and are therefore more likely to be exposed to the associated hazards as well as benefits. In this chapter, we discuss water-borne infectious diseases and their transmission as determinants of human health within a wetland setting. For a discussion of the benefits wetlands provide to human health and well-being, see Horwitz et al. (2012), and for vector-borne infectious diseases associated with standing water, see Carver et al. (2015) in this publication.

Fig. 1 Family living on Tonlé Sap Lake, Cambodia. Photo: Nelson Lau

Infectious Diseases in Wetlands

The biodiversity found in wetlands are largely responsible for the important ecosystem services they provide. However, 'biodiversity' also encompasses organisms that can be detrimental to human health.

Wetlands provide an aquatic or semi-aquatic, often relatively stable, environment with optimal conditions (in terms of water flow, nutrients, temperature, salinity) for the survival or proliferation of certain bacteria, protozoa, viruses, and helminths, as well as their plant and animal hosts, reservoirs or vectors. A subset of these organisms can cause disease in humans, some of which are associated with significant disease burden.

This chapter focuses on water-borne infectious diseases affecting humans, where the causative organisms are transmitted in aquatic or semi-aquatic (e.g. water-logged soil) environments inherent to wetlands. The transmission of water-borne pathogens can occur via the faecal-oral pathway, or through skin or mucous membrane surfaces. Though inland wetland environments alone cover millions of hectares on all inhabited continents (see Finlayson and D'Cruz 2005), and water-borne diseases are a threat to people everywhere (see Tables 1 and 2), the burden of water-borne diseases is heavily skewed by regional socio-economic status. Water-borne diseases (also occurring outside of the wetland setting) are a major cause of mortality and illness amongst the world's poorest populations. For example, the World Health Organisation estimates that, globally, 4 billion cases of diarrhoea per year arise from lack of access to clean water (Finlayson and D'Cruz 2005). In 2004, there were 1.9 million estimated deaths worldwide due to diarrhoea and other diseases associated with poor water and sanitation (WHO 2009c).

Diseases Transmitted via the Faecal-Oral Route

Faecal-oral disease transmission occurs when pathogens contained in the faeces of a host animal are introduced to the gastrointestinal tract of another host. Within a wetland setting, human exposure to pathogens transmitted by the faecal-oral route occurs through drinking contaminated water, incidentally ingesting contaminated water through swimming/washing, or consuming contaminated food that originated from wetlands (e.g. vegetables, molluscs and fish). Contamination of water occurs by faecal matter being excreted or washed into the wetlands (e.g. during heavy rainfall or flooding). Infected people, wild animals and livestock are all important sources of contamination. Land use and other anthropogenic activities that maximise the presence of hosts, and excretion of their faecal matter into wetlands, therefore increase infection risk. Common examples of faecal-oral diseases transmitted in wetlands are shown in Table 1.

Table 1 Examples of infectious diseases transmitted by the faecal oral route resulting from human exposure to wetland environments

Disease & Causative Organism	Burden & distribution	Ecology	Risk factors (Wetland related)	Outbreak example
Enterohaemorrhagic E.coli poisoning caused by *Escherichia coli* Pathogenic strains such as O157 and Enterotoxigenic *E.coli* (ETEC) (Bacterium)	Common Worldwide ETEC most common in developing countries, with an estimated 200 million cases per year (WHO 2009b)	Found in the gut of warm-blooded animals, commonly cattle	Proximity of agricultural areas to wetlands used for drinking, swimming, food	Lake in Oregon, USA, Summer of 1991: 21 children developed bloody diarrhoea and haemoloytic-ureamic syndrome from *E.coli* 0157 after swimming in a county lake. Most likely caused by faecal contamination by swimmers (Keene et al. 1994)
Campylobacteriosis caused by *Campylobacter* spp. (Usually *C. jejuni* & *C.coli*) (Bacterium)	Common worldwide, with an estimated 400 million cases per year (WHO 2009a)	Found in the gut of warm-blooded animals, commonly poultry & livestock	Proximity of agricultural areas to wetlands used for drinking, swimming, food	Stream near Utti, Finland, July 1987: 75 young men developed diarrhoea, abdominal pain and fever after drinking untreated water from a cold clear stream in an uninhabited area. Ducks from a swamp upstream were identified as the likely source (Aho et al. 1989)
Cholera (see Case Study 1 caused by *Vibrio cholerae* O1 or O139 (Bacterium)	Estimated 3–5 million cases per year in tropical regions, predominantly in developing countries (WHO 2013a)	Occurs in brackish and saltwater environments, often in association with zooplankton, algae, plants and possibly avian hosts. Can survive and multiply independently of human hosts	Overcrowding, poor waste sanitation and drinking water treatment Warmer temperatures, algal and zooplankton blooms, certain aquatic plants (e.g. water hyacinth)	Seasonal epidemics around Bay of Bengal (Kolkata, India & Matlab, Bangladesh), 1998–2006. Hundreds of people relying on untreated river water for drinking and washing suffered from acute diarrhoea. Environmental factors significantly associated with case numbers were the El Niño event of 1998, driving increased algal levels, and rainfall anomalies (Constantin de Magny et al. 2008)

Table 1 (continued)

Disease & Causative Organism	Burden & distribution	Ecology	Risk factors (Wetland related)	Outbreak example
Typhoid and Paratyphoid fever caused by *Salmonella typhi* & *S. paratyphi* (Bacterium)	Estimated 17 million cases per year worldwide, predominantly in developing countries (WHO 2001a)	Contamination of water occurs by human faeces	Overcrowding, poor management of waste and sanitation, and unclean drinking water	Deir az-Zor Syria, February 2013. 2500 people suffered from diarrhoea after drinking untreated, sewage contaminated water from the Euphrates River as lack of fuel and power compromised access to safer groundwater (Anonymous. 2013)
Cryptosporidiosis (see Case Study 3) caused by *Cryptosporidium* spp. (Protozoan)	Common worldwide	Livestock (cattle, sheep, goats) are the most common zoonotic reservoirs	Proximity of water bodies to agricultural areas Recreational activities such as swimming, drinking untreated water whilst camping Heavy rainfall and flooding	Milwaukee, Wisconsin, USA. March-April 1993. 403,000 cases of gastrointestinal illness and 100 deaths after the city water treatment failed to remove Cryptosporidium oocytes from the turbid water supply. Water supplying the plant from Lake Michigan may have been contaminated by upstream cattle pastures, slaughterhouses or human sewage, and exacerbated by spring rains and snow runoff (MacKenzie et al. 1994)
Giardiasis (see Case Study 3) caused by *Giardia lamblia* (Protozoan)	Common worldwide	Zoonotic reservoirs include wild animals (e.g. beavers and birds) and domestic animals (e.g. dogs, cats, cattle)	Proximity of water bodies to wildlife Recreational activities such as swimming, drinking untreated water whilst camping Heavy rainfall and flooding	Red Lodge, Montana, USA, summer of 1980. 780 people suffered from gastrointestinal illness, preceded by ash fall from an eruption of Mt St Helen volcano and above average temperatures, causing significant snow runoff into and water turbidity of the creek supplying the town water supply, which was not filtered. Potential sources of contamination within the watershed included overflowing septic tanks, hikers, beavers and domestic animals (Weniger et al. 1983)

Table 1 (continued)

Disease & Causative Organism	Burden & distribution	Ecology	Risk factors (Wetland related)	Outbreak example
Hepatitis A & E Hepatitis A caused by Virus (HAV) and Hepatitis E Virus (HEV) (Virus)	Found worldwide, most commonly in developing countries 1.4 million estimated cases of Hepatitis A per year 20 million estimated cases of Hepatitis E per year (WHO 2001b)	Hepatitis A seen only in humans. Pigs and rats are known reservoirs of Hepatitis E Hepatitis A can persist in estuarine environments	Poor waste sanitation Shellfish cultivated in waters polluted with faecal matter Proximity of water sources to piggeries	Wallis Lake, New South Wales, Australia, early 1997. Over 444 people suffered from Hepatitis A (with ~14 % hospitalised and one death) after eating oysters farmed in the lake. Contamination attributed to nearby sewage outlets flowing into the water system, with above average rainfall 2–4 months before the outbreak (Conaty et al. 2000)
Noroviruses caused by strains of Norwalk Virus (Virus)	Common worldwide-most common cause of winter gastroenteritis in industrialised countries	Humans are the reservoir for Noroviruses, though domestic dogs may transmit it (Summa et al. 2012) Noroviruses can persist in estuarine environments	Poor waste management, sanitation and water treatment Shellfish cultivated in waters polluted with faecal matter	West Gotland, Sweden, 2008. Over 300 cases of gastroenteritis caused by Noroviruses in recreational swimmers using two lakes. Mussels in a nearby archipelago, exposed to a sewage effluent plume (thought to be exacerbated from preceding heavy rains) were found to be contaminated with the same Norovirus strains as the swimmers (Nenonen et al. 2008)

Diseases Transmitted via Skin or Mucous Membrane Contact

Certain pathogens infect humans by directly penetrating skin surfaces (particularly through scratches, sores, cuts, and penetrating injuries), or mucous membranes (e.g. eyes, nostrils). Human exposure to these pathogens usually occurs through direct contact with water and/or soil. Activities involving partial or full immersion such as swimming, washing, and fishing, are therefore commonly associated with exposure. Like most faecal-oral pathogens, many skin-transmitted pathogens originate from animal reservoirs and their presence and abundance in the environment depends on the shedding of pathogens by reservoirs into the water. Some parasites, such as the flatworms causing schistosomiasis, require certain species of freshwater snail as intermediate hosts (Case Study 2). Unlike passively acquired water-borne infections, such pathogens actively seek out humans and other warm blooded animals as definitive hosts. Completion of their lifecycle thus requires the presence of both intermediate and definitive hosts species within the wetland. Common examples of skin and/or mucous membrane transmitted diseases associated with wetlands are listed in Table 2.

Environmental and Anthropogenic Drivers of Infectious Disease Transmission in Wetlands

Environmental change and disruption, in its many forms and sources, disrupts the natural equilibrium existing between and within abiotic (non-living) and biotic (living) components of a wetland. Changes in the presence, abundance and transmission dynamics of pathogens may result from such disturbances (Patz et al. 2000), altering exposure risk and disease burden in humans. Both the complexity and our lack of understanding of these dynamics mean that changes in infectious disease risk in wetlands are often difficult to predict. Furthermore, wetlands ecosystems are not only vulnerable to internal, spatially immediate perturbations, but often have a high level of connectivity to adjacent landscapes via watershed systems, thus increasing vulnerability to environmental change.

Environmental change leading to altered infectious disease dynamics is categorised and discussed below under three main themes: climate, human activities, and biodiversity. These types of change are highly interrelated, and can interact synergistically as different links of the causal chain leading to disease outbreak in a wetland setting. Figure 2 illustrates how the transmission of infectious diseases can be driven by the complex interactions between human activities, climate, and other biotic and abiotic environmental factors.

Table 2 Examples of infectious diseases transmitted through skin or mucous membrane contact that arise from human exposure to wetland environments

Disease & Causative Organism	Burden & distribution	Ecology	Wetland-related risk factors	Emergence/outbreak example
Leptospirosis (see Case Study 4) caused by *Leptospira* spp. (Bacterium)	Common worldwide, particularly in tropical environments and developing countries	Found in soil and water, various mammal species act as reservoirs	Proximity to livestock, wild mammals and rats, and agricultural areas (e.g. rice paddies) Heavy rainfall and flooding Poor waste management Recreational activities such as swimming	Increasing incidence in Kolenchery, Kerala, India- a landscape that consists of marshy and dry land interspersed by rivulets and ponds. Irrigation of dry land for cultivation during summer started in the mid 1980s Since 1987 a yearly increase in leptospirosis incidence has been observed (Kuriakose et al. 1997)
Melioidosis caused by *Burkholderia pseudomallei* (Bacterium)	Endemic to Southeast Asia and Northern Australia (WHO 2013b)	Found in soil and surface water Also affects animals but not known to be zoonotic	Exposure to contaminated soil and water. Most prevalent during and after rain periods	Southern Taiwan, July-September 2005- 40 confirmed cases of melioidosis following widespread flooding caused by a typhoon (Ko et al. 2007)
Schistosomiasis (see Case Study 2) caused by *Schistosoma* spp. (Helminth)	Tropical regions- mainly Africa but also Southeast Asia, South America, Middle East, the Caribbean. In 2011, an estimated 28.1 million people were treated for schistosomiasis (WHO 2013c)	Requires fresh water snail (host species varies with parasite species) as intermediate host for asexually reproducing, larval stage. Warm-blooded vertebrates are the definitive hosts	Dams & canals associated with agricultural irrigation systems Mining & other human activities	Central Sudan, 1970. Intestinal schistosomiasis more prevalent throughout the Gezira irrigation system after the addition of a storage dam (Amin et al. 1982)

Table 2 (continued)

Disease & Causative Organism	Burden & distribution	Ecology	Wetland-related risk factors	Emergence/outbreak example
Buruli Ulcer caused by *Mycobacteria ulcerans* (Bacterium)	Found in over 30 countries across Africa, the Americas, Australia, Asia and the West Pacific. 5000-6000 reported cases p.a. (WHO 2012a)	Mode of transmission unclear, probably varies with epidemiological and geographical setting (see (Merritt et al. 2010) for review). Skin contact with environment suspected, as are aerosol and vector-borne transmission (by mosquitoes or aquatic insects). Aquatic insects, fish, freshwater snails, possums are likely reservoirs	Contact with or proximity to aquatic environments which have undergone environmental disturbance from flooding, damming, construction, deforestation, agriculture activities, irrigation systems, mining	Togo near the Benin border, November 1994. Two separate cases featuring massive skin ulcers in children living in villages proximate to river systems. Cases were attributed to environmental exposure to river systems which flooded 4–5 months previously (Meyers et al. 1996)
Spam disease caused by *Mycobacteria marinum* (Bacterium)	Relatively uncommon, known to occur in aquarium keepers	Certain fish species as reservoirs	Contact with stagnant water bodies. Recreational and occupational activities such as swimming, farming, and fishing	Satawonese Islands, Central Pacific. 39 case patients examined showed chronic, progressive verrucous or keloidal plaques. All were taro farmers, and contact with water filled bomb craters (from World War II) was a major risk factor (Lillis et al. 2009)

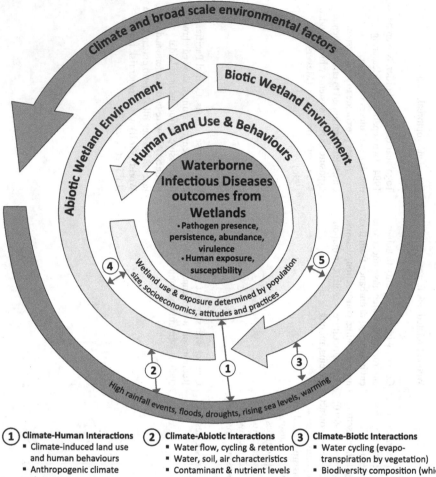

Fig. 2 Interaction of the major factors affecting infectious disease risk to humans in a wetland setting

Climate

Climate has been known to affect water-borne diseases principally by warming temperatures and by heavy rainfall and flood events (Rose et al. 2001; Hunter 2003), both of which affect wetland environments. On a global scale, high rainfall and floods, as well as the warmer temperatures which often drive them, are predicted to increase in frequency and intensity in the future (Pachauri and Reisinger 2007).

Rainfall and Flooding

There is a strong correlation between high rainfall events, which may culminate in flooding (though flooding can also result from tidal surges or snowmelts), and outbreaks of various infectious water-borne diseases. A review by Curriero et al. (2001) of water-borne disease outbreaks in the USA found that 68% of the outbreaks reported to the Environmental Protection Agency between 1948 and 1994 were preceded by heavy rainfall. Heavy rainfall and flooding could drive water-borne infectious disease transmission through a number of mechanisms that relate both to the presence and persistence of pathogens, and to subsequent human exposure.

Increased precipitation leads to swollen and often altered water flow routes, linking upstream sources of pathogen contamination or pollution into wetlands (Hunter 2003). For example, runoff from livestock pastures may carry faecal matter into surface water bodies such as streams and lakes, causing faecal-oral transmission of disease to people using the wetlands downstream. Concentrations of faecal indicator organisms during an unusually wet winter caused by El Niño climatic patterns were found to be several times higher than at any other time of the year in Charlotte Harbour Florida (Lipp et al. 2001). Once pathogens are present, above average rainfall can also create optimal conditions within a wetland environment for proliferation, such as changes in nutrient levels, water flow, turbidity, salinity, pH, and other aspects of water composition. Algal blooms fuelled by both high nutrient levels and by the lowered salinity caused by freshwater runoff into estuaries have caused counts of pathogenic *Vibrio* bacteria to peak on various occasions in Florida (Rose et al. 2001).

Increased water turbidity caused by run off also can place major stress on the filtration systems of water treatment plants (Hunter 2003). The infamous cryptosporidiosis outbreak of Milwaukee, Wisconsin in 1993 was partly attributed to the city water treatment system being compromised by sediment run off caused by heavy rainfall (MacKenzie et al. 1994; Rose et al. 2001). Floods may more directly increase human exposure to water-borne diseases by washing pathogens from their natural environment (e.g. a wetland) into urban areas where contact with people is highly likely (see Fig. 3). This scenario frequently occurs in low-lying, densely populated areas with poor sanitation, such as urban slums in lower income countries. The numerous outbreaks of leptospirosis following flooding provide an example of this problem (see Lau et al. 2010 and Case Study 4).

Warming

Summer is commonly the season in which infectious diseases from aquatic environments peak in high-income regions such as North America (Rose et al. 2001;

Fig. 3 Flood-prone community on Saibai Island, Torres Strait Islands, Australia. (Photo: Colleen Lau)

Dziuban et al. 2006). Warmer water temperature has been shown to be optimal for the survival and proliferation certain water-borne pathogens. This may be due to the direct effects of temperature on the pathogens, or the effect temperature has on the abundance of host or associated species.

Higher rates of isolation of *Vibrio cholerae* from the environment in warmer months have been reported in North America (Lipp et al. 2002). Warmer temperatures are known to drive phytoalgal blooms in shallow coastal areas, which are followed by zooplankton blooms, and closely associated *Vibrio cholera* (Lipp et al. 2002) (see Case Study 1, below). The parasitic amoeba *Naegleria fowleri*, an agent of meningoencephalitis, has also been shown to occur in thermally heated-water bodies (Sykora et al. 1983). Conversely, warmer water temperatures may reduce the survival of *Vibrio cholera* in some cases, and also of *E.coli, Campylobacter* spp. and enteroviruses (Hunter 2003).

Importantly, warming also increases the risk of infection by influencing human behaviour and exposure to wetlands, e.g. by encouraging water-based recreational activities. Many outbreak reports in high-income countries countries involve swimmers rather than contamination of drinking water (e.g. Dziuban et al. 2006).

It has also been proposed that increased evaporation rates or drought driven by warming can drive the transmission of zoonotic water-borne diseases such as leptospirosis (Dufour et al. 2008; Lau et al. 2010). Reduced size and availability of water bodies may cause crowding of both reservoir species and humans, and consequently lead to increased pathogen load in the water, more contact between animals and humans, and greater opportunities for transmission. Pathogen concentration could also be higher when water levels are low.

Like the flooding of important wetlands and water sources, the drying of wetlands in drought can cause the forced displacement of populations. In low-income countries, this is likely to lead to poorer living conditions with overcrowding, inadequate clean water and sanitation, and slum-like conditions. Such environments provide favourable conditions for the proliferation and transmission of pathogens. Combined with the vulnerability and poor resilience typical of displaced populations, infection risk and disease burden are likely to be high.

Case Study 1

Cholera: A faecal-oral water-borne diarrhoeal disease with a complex ecology

Each year, an estimated 3–5 million people worldwide develop cholera by ingestion of water or food contaminated with the bacterium *Vibrio cholerae* (toxogenic strains O1 and O139 are the cause of epidemic cholera), resulting in an estimated 100,000–120,000 deaths (WHO 2013a). Cholera is an acute intestinal infection which is asymptomatic in approximately 75% of cases, but severe cases typically result in profuse watery diarrhoea that can rapidly lead to severe dehydration and death (WHO 2013a). The vast majority of cases occur in developing countries, where water treatment and sanitation infrastructure are non-existent or inadequate. *Vibrio cholerae* bacteria have been detected in coastal, estuarine and riverine waters throughout the world, though the disease is only endemic in certain tropical and subtropical regions (Lipp et al. 2002). Epidemic foci in the past few decades originated from South Asia and spread to sub-Saharan Africa (95% of cases reported between 2001 and 2009 were from Africa (WHO 2013a)) and South and Central America. Over 60% of cases reported globally for 2011 occurred in Haiti (WHO 2012b).

The transmission dynamics of *Vibrio cholerae* exemplifies the intricate interplay between the biotic and abiotic conditions of the wetland environment; the landscape, climatic and human factors shaping that environment; and the human populations at risk of infection. This spatially and temporally variable relationship between the disease, the aquatic environment and climate has been termed the "Cholera paradigm" (Colwell 1996), and is a focus of ongoing research. The complexity of cholera transmission dynamics are partly attributable to the fact that, while *Vibrio cholerae* are capable of existing independently within an aquatic environment, they are most often attached to tiny invertebrate species such as copepods that make up zooplankton (Constantin de Magny et al. 2008). One major factor affecting pathogen

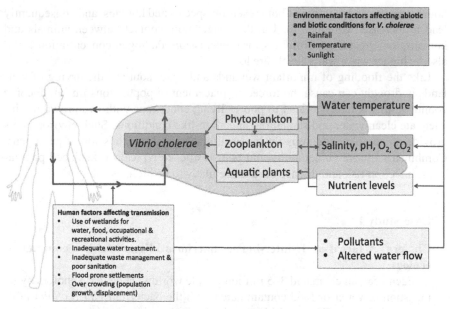

Fig. 4 The principal known components of drivers of *Vibrio cholerae* within a wetland environment and its transmission to humans as the cause of cholera

abundance is therefore biotic, which in turn is influenced by a multitude of factors, such as the phytoplankton (made up of algae) which zooplankton feed on. Cholera epidemics are often preceded by algal blooms (Constantin de Magny et al. 2008), which are driven by proximate abiotic factors related to water chemistry and temperature resulting from broader scale climatic and environmental conditions such as rainfall and nutrient influx (see Fig. 4, and Lipp et al. 2002). For example, in coastal, cholera-endemic areas of the Bay of Bengal, temporal and spatial analysis of environmental data found a positive correlation between phytoplankton levels (driving cholera outbreaks) and periods of high river discharge of terrestrial nutrients. Environmental conditions may not only affect *V. cholerae* abundance, but also the prevailing serotype, and even the expression of virulence genes (Lipp et al. 2002), both of which have direct impact on the risk and severity of infections.

In the great lakes region of Africa (including parts of Burundi, Rwanda, Democratic Republic of Congo, Tanzania, Uganda and Kenya), cholera has reemerged and persisted since 1977, with over 322,000 cases reported between 1999 and 2008 (Nkoko et al. 2011). Several studies have found that living on a lake or river shore is statistically correlated with cholera infection (Nkoko et al. 2011), with many residents using sewage-contaminated lake water for drinking and cooking. Furthermore, Nkoko et al. (2011) found that lakeside locations were the only areas where the disease persisted continuously over the 30-year study period, suggesting these lakes are the environmental reservoir for the region. The dry season increase of cholera cases are likely attributable to fishermen who camp on seasonally exposed islands in the lake in crowded conditions, with no access to clean water or sewage systems. The fishermen thus become human reservoirs for cholera and perpetuate the transmission cycle, even if the pathogen would otherwise have disappeared from the environment (Nkoko et al. 2011).

Human Activities

Land use changes arising from human activities are the primary drivers of many infectious disease outbreaks and also of the geographic distribution of endemic disease (Patz et al. 2000). Human activities related to urbanisation, agriculture, mining or industry frequently cause changes in both hydrology and biodiversity, leading to alterations of wetlands and their disease dynamics. Globally, agriculture alone occupies half of all land and consumes over two thirds of the world's fresh water (Horrigan et al. 2002), demonstrating the extent of anthropogenic impact on the earth's ecosystems.

The direct alteration and pollution of wetlands and their watersheds are two results of human activities that drive the presence and proliferation of pathogens in wetlands, whilst human behaviour determines exposure to these pathogens.

Direct Alteration of Watercourses, Watersheds and Wetlands

Most human activities within and around wetlands cause changes to watercourses, including the source, direction, volume and velocity of water input and output, and their nutrient and pathogen load. Activities such as damming, draining and irrigation may even create a new wetland or remove it completely. Similar to the manner in which rainfall could drive disease transmission in wetlands, altered water flow may link sources of pollution, such as sewage or agricultural pastures to wetlands. Contaminated water can also be transported from wetlands to susceptible human populations through altered watercourses. Altered environments may also drive disease by creating optimal environmental conditions for pathogens and/or their hosts, as in the case of schistosomiasis (see Case Study 2).

Case Study 2

Schistosomiasis—A parasitic water-borne disease

Schistosomiasis (also known as Bilharzia) is caused by infection of humans by certain species of the trematode flatworm genus *Schistosoma*. Malaria is the only parasitic disease affecting more people worldwide. As of mid 2003, an estimated 779 million people were at risk of schistosomiasis,

and 207 million were infected (Steinmann et al. 2006). In 2011, it was estimated that at least 243 million people required treatment for the disease
(WHO 2013c). Common symptoms of acute infection (Katayama fever)
include rash, itching, fever, chills, coughs, and muscle aches; while chronic
infection results in inflammation and scarring of the intestines and bladder.
In children, schistosomiasis can result in malnutrition, anemia and cognitive
impairment (CDC 2012).

Schistosomiasis transmission has been recorded in 78 countries throughout tropical and subtropical regions of Africa, Asia, South America and the
Caribbean (WHO 2013c). Each pathogenic *Schistosoma* species has a different geographic distribution, largely dependent on that of the corresponding freshwater snail species serving as intermediate hosts (Patz et al. 2000;
DeJong et al. 2001). Despite this trans-continental distribution, 90 % of people requiring treatment in 2011 lived in Africa (WHO 2013c).

The *Schistosoma* lifecycle (summarised in Fig. 5) requires a freshwater
environment, with access to both aquatic snail hosts and warm-blooded ver-
tebrate hosts, i.e. birds, humans, or other mammals. People spending time
in natural and artificial freshwater wetlands containing the parasite and its
snail host species risk infection by free-swimming *Schistosoma* larvae (cer-
cariae), which actively seek hosts and penetrate their skin. In the body, larvae
mature in the portal veins before paired adults migrate to the bowel, bladder,
or female genital tract depending on the species. Eggs pass into these tracts
and are shed through urine and faeces (CDC 2012). If eggs are shed into a
wetland environment (commonly by urinating directly into the water, or rainfall washing nearby faeces into the wetland), they hatch in the water and the
miracidia larvae infect snail hosts where asexual reproduction occurs.

Water temperature, flow velocity, as well as rainfall (which may be needed
to wash egg bearing host faeces into the water), are all important abiotic factors determining wetland suitability for *Schistosoma* spp. in their free swimming stages (Appleton and Madsen 2012). However, as obligate parasites,
host availability is crucial. For the snail hosts (*Bulinus spp.*) of *Schistosoma
masoni* and *S. haematobium* common to Southern Africa, water temperature,
flow velocity, chemistry (e.g. dissolved oxygen, calcium carbonate), turbidity, substrate and habitat structure are some of the main factors that determine
distribution (Appleton and Madsen 2012). Perennial water bodies are generally more suitable for these snail hosts than seasonal wetlands (Appleton and
Madsen 2012). Snail hosts have commonly colonized wetlands created or
altered by humans (e.g. dams or irrigation systems), where the slowed water
flow provides suitable conditions for aquatic weeds (e.g. water hyancinth) to
proliferate, creating predator- and competitor-free refuges for snails to breed
in (Appleton and Madsen 2012). Such environmental changes have resulted
in local emergence or increasing incidence of human schistosomiasis (Amin
et al. 1982; N'goran et al. 1997; Patz et al. 2000; Steinmann et al. 2006;
Appleton and Madsen 2012).

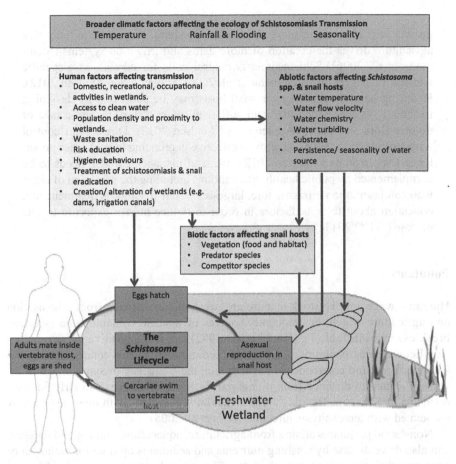

Fig. 5 Human and environmental factors affecting schistosomiasis transmission within a wetland environment

For people using wetlands, swimming, fetching water, washing clothes, fishing and rice cultivation are common activities associated with exposure to *Schistosoma* parasites (Appleton and Madsen 2012; WHO 2013c). Prevalence tends to be higher in children due to increased time spent immersed in water at hot times of the day (when the free swimming cecariae are most active), and a tendency to urinate whilst in the water, thus re-infecting popular swimming locations (Steinmann et al. 2006; Appleton and Madsen 2012). Travellers engaging in ecotourism activities such as swimming or kayaking in schistosomiasis-endemic areas are also at risk (Schwartz et al. 2005).

Without preventative measures, schistosomiasis incidence is likely to increase in the future. Population growth will result in higher density populations living in closer proximity to wetlands, therefore increasing contamination of wetlands with human waste (Appleton and Madsen 2012). Population growth will also create more *Schistosoma*-endemic wetlands, as intensive agriculture drives the creation of more dams and irrigation systems (Stein-mann et al. 2006). Schistosomiasis control calls for careful, case-specific wetland management (Steinmann et al. 2006; Appleton and Madsen 2012). Reducing habitat suitability for snail hosts may be addressed by designing irrigation systems which maintain sufficiently fast waterflow and a host of diverse flora and fauna (Appleton and Madsen 2012). Direct snail control with molluscicides have also proven effective in reducing local disease preva-lence (Appleton and Madsen 2012). Wetland management also needs to be complemented by public health interventions including the provision of clean water and sanitation infrastructure, large-scale schistosomiasis treatment, and education about the risk factors in order to reduce human exposure to this disease (WHO 2013c).

Pollutants

Human sewage and livestock manure are two pollutants arising from urbanisation and agriculture which are frequently sources of wetland contamination (see out-break examples in Table 1). Cabelli et al. (1982) reported a positive linear relation-ship between gastroenteritis in swimmers and microorganisms counts in sea water (marine *E.coli* and enterococci). Furthermore, the frequency of symptoms was in-versely related to distance from municipal waste water source. Similarly, in slum residents of Salvador, Brazil, living in flood-prone valleys with open sewers was associated with leptospirosis infection (Reis et al. 2008).

Non-faecal pollutants arising from agriculture, aquaculture, mining and industry can also drive disease by washing nutrients and sediments optimal for pathogen or host growth into wetland environments. The direct links between eutrophication (excessive plant or microorganism growth, such as algal blooms, caused by nutri-ent inputs) and cholera outbreaks exemplify the effect of nitrogen and phosphorus inputs into aquatic environments (Epstein 1993).

Case Study 3

Cyrptosporidiosis and Giardiasis: faecal-oral infections arising from wetland or watershed contamination

Giardiasis and cryptosporidiosis are intestinal infections caused by the unicellular protozoan parasites *Giardia intestinalis* (also known as *G. duode-nalis* and *G. lamblia*), and members of the *Cryptosporidium* genus, primarily *C. parvum* and *C. hominis*, respectively. Both infections can cause acute diar-rhoea, abdominal cramps, and nausea. Symptoms can last for many weeks if untreated, or recur after treatment. In children, chronic giardiasis can lead to malabsorption, weight loss, and delays in physical and mental development.

Cryptosporidiosis may cause serious or even fatal complications in immu-nocompromised patients (CDC 2013b). Cryptosporidiosis and giardiasis occur in both high and low-income countries throughout the world. An estimated 33 % of people in developing nations have had giardiasis (CDC 2013c), while the prevalence of *C. parvum* in stool samples of gastroenteritis patients was reported to be as high as 20 % in Africa, Asia, Australia, and south and central America (Current and Garcia 1991; Fricker et al. 2013).

Giardia and *Cryptosporidium spp.* possess key characteristics that increase their ecological success as water-borne pathogens likely to infect people exposed to freshwater wetland environments (Weinstein et al. 2000). In addition to having human reservoirs, both pathogens are potentially zoonotic. Certain strains of *C. parvum* responsible for human outbreaks have cattle hosts. *G. intestinalis* is prevalent in a wide range of livestock (including cattle and sheep) and wild mammals, and at least one genotype has been found to infect both humans and other mammals (Monis and Thompson 2003). Animal and human faeces thus provide sources of contamination of wetland waters. Furthermore, *Giardia* cysts or *Cryptosporidium oocysts* shed through faeces into the environment are robust and capable of persisting in water and soil (Mawdsley et al. 1995). Oocysts and cysts can therefore wash across agricultural pastures, through watersheds, or through sewage systems, into surface waters and persist there (Hansen and Ongerth 1991) (see Fig. 6). In environmental surveys across North America, *Giardia* and *Cryptosporidium* were found more frequently in water bodies receiving sewage and agricultural discharge than in pristine waters (Rose et al. 1991). Importantly, *Cryptosporidium* oocysts are resistant to chlorination and most filtration measures employed in water treatment for drinking water supplies (Weinstein et al. 2000).

Fig. 6 Livestock pastures near the Rakaia River, New Zealand. Photo: Ghislaine Arteon

Epidemiological studies repeatedly demonstrate that human use of wetlands, whether as a source of drinking water or for recreational activities, could expose people to significant risk of cryptosporidiosis and giardiasis. During a summer cryptosporidiosis outbreak of some 2000 persons visiting a New Jersey state park, swimming in a lake was strongly associated with infection (Kramer et al. 1998). In New England, USA, giardiasis was significantly associated with drinking untreated surface water and swimming in fresh water (Dennis et al. 1993). Sporadic giardiasis in South West England was associated with swallowing water while swimming, and recreational fresh water contact (Stuart et al. 2003). Groups most likely to drink untreated or inadequately treated surface water include populations in rural and/or developing areas (both residents and travellers), and recreational hikers and campers.

The ability of pathogenic *Cryptosporidium spp.* and *Giardia intestinalis* to contaminate a wetland environment from sources not immediately proximate to that wetland emphasises the importance of risk mitigation at a broader, ecological level (Weinstein et al. 2000). Considering hydrological drainage patterns, reducing agricultural runoff and careful planning of farm and pasture practices accordingly are possible approaches (Weinstein et al. 2000, Mohammed and Wade 2009). Pre-emptive management is particularly important for minimising *Cryptosporidium spp* contamination drinking water sources, since the pathogen is difficult to eliminate during water treatment.

Human Use of Wetlands

Human interactions with wetland environments arising from food and water provision, occupational or recreational and cultural activities vary greatly between individuals, communities and wider populations. Consequently, their exposure to infectious disease agents found in wetlands will also vary. Socio-economic status and level of water and sewage infrastructure profoundly influence the reliance on and exposure to wetlands and their pathogens. Populations in low-income regions where people must collect water directly from wetlands for drinking, domestic, and livestock purposes have a much higher water-borne disease burden than urban populations from developed countries who drink high quality treated water and largely use wetlands for non-essential recreational pursuits. In 2004, for example, some 1.9 million deaths resulted from unsafe water, sanitation and hygiene issues globally, with less than 0.5% of these deaths coming from high-income countries (WHO 2009c).

Education level and knowledge are another important factor influencing human disease risk in wetland environments. The risk levels individuals and communities expose themselves to can be greatly reduced by knowledge of risk factors and risk mitigation strategies (e.g. hand washing between exposure to water and eating as a precaution against faecal-oral infection). In American Samoa, knowledge about

Fig. 7 A public health warning of leptospirosis risk in streams in American Samoa. Photo: Nelson Lau

leptospirosis was found to be associated with a lower risk of infection, probably because of precautions taken to reduce exposure, e.g. wearing protective clothing and gloves and avoiding flood waters (Lau et al. 2012) (see Fig. 7). At a community or regional scale, knowledge of how preservation and appropriate management of ecosystems such as wetlands benefits human well-being can lead to less environmental degradation and better health outcomes, including lowered disease risks. Similarly, prevailing attitudes (shaped by culture, history, competing economic/social interests) at individual, community and regional levels towards wetland use, preservation and management, influence disease risk and wetland integrity. The previously widespread practice of draining wetlands to reduce mosquito populations and malaria risk is an example of how attitudes and knowledge shaped both disease risk and the fate of an ecosystem (Horwitz and Finlayson 2011). This example also raises the point that human attitudes and behaviours concerning use of wetlands are dynamic over time and space, and prone to change in response to changes in the abiotic and biotic environment.

Children are particularly vulnerable to water-borne infectious diseases in wetlands because of their behaviour, e.g. poor personal hygiene and hand washing, increased time spent outdoors, and playing in water. For many pathogens, children are more susceptible to infection because of their naïve immune systems and more likely to develop serious complications, e.g. severe dehydration from cholera (Lau and Weinstein 2011).

Biodiversity Loss

Despite high biodiversity having the potential to increase infectious disease exposure and transmission to humans, a substantive body of evidence from a variety
of settings support the notion that biodiversity loss generally results in increased
infectious disease transmission (Keesing et al. 2010). Wetland biodiversity can affect disease risk to humans by regulating pathogen numbers, either directly or by
regulating hosts.

Bioregulation occurs when other species in the eco-system prevent a pathogen
or host species from establishing or becoming overly abundant; whether by preying
on it or competing with it for resources (Ostfeld and Holt 2004). More biodiverse
ecosystems are more likely to support predator and competitor species that exert
effective bioregulation on disease causing species (Ostfeld and Keesing 2000). The
bioregulation concept is illustrated by the human overfishing of snail-eating fish
in Lake Malawi, which was followed by a population boom in a snail host species
of *Schistosoma haematobium*, and increased local prevalence of schistosomiasis
(Madsen et al. 2001).

The dilution effect acts when a pathogen has a number of species it can use as
a host or reservoir within an ecosystem. If these different species are differentially
competent hosts, and the pathogen has a finite number of transmission opportunities, then transmission to competent hosts are 'diluted' by transmission events being
'wasted' on less competent hosts, reducing overall disease transmission (Ostfeld
and Keesing 2000). A recent study examining *Ribera ondatrae* infection (flatworm
parasite of amphibians) across 345 wetlands in the US found that highly competent
host species dominated less biodiverse wetlands. Consequently, pathogen transmission and pathology was significantly lower in species rich wetlands than in species
poor wetlands (Johnson et al. 2013).

Loss of wetland vegetation can also directly increase exposure to pathogens,
as has been shown for campylobacteriosis in New Zealand (Weinstein and Woodward 2005). Removal of deep rooted vegetation firstly results in soil erosion and
increased runoff from surrounding areas into wetlands, therefore making contamination from sources such as livestock pastures more likely. Importantly, the viability of pathogens in wetlands can also increase when pathogens which have not yet
exceeded their half life fail to be sequestered by natural vegetation (Weinstein and
Woodward 2005). Removal of vegetation can also create conditions ideal for pathogen growth, as in the case of *Mycobacterium ulcerans,* which favours increased
water temperatures, sediment and nutrient levels arising from loss of riparian vegetation (Merritt et al. 2010).

Case Study 4

**Leptospirosis: A Water-borne Disease and its Environmental Drivers of
Transmission in a Wetland Setting.**
    Leptospirosis is the most common bacterial zoonotic disease, with an esti-
mated 500,000 severe cases occurring worldwide each year. While leptospirosis

can be asymptomatic or manifest as a mild non-specific febrile illness, it can also cause serious complications such as renal and liver failure, pulmonary haemorrhage, meningoencephalitis, and death (WHO 2011).

Leptospires rely on various mammal species as maintenance hosts (including rats, pigs, cattle, horses, dogs, marsupials and bats), and can survive in aquatic and humid environments once shed via animal urine. Human infection occurs through mucosal or skin surfaces upon contact with infected animal tissue or urine, or through contaminated water and soil. Occupational and recreational exposures related to wetlands, such as rice harvesting, fishing, swimming, and kayaking are therefore important risk factors. Although developing countries in tropical regions face the highest disease burden, leptospirosis outbreaks occur in a range of settings—differing in climate, level of urbanisation, and socio-economic development.

Animals and the environment are important factors that determine the transmission ecology of leptospirosis, and infection risk is therefore intimately linked to the interplay between the environment (including anthropogenic land use), animal host species, and human behaviour (see Fig. 8). Consequently, leptospirosis provides a pertinent example of an infectious disease where transmission is driven by environmental change and disruption of wetlands.

The 2002 outbreak that affected hundreds of people in Jakarta, Indonesia (Victoriano et al. 2009) is one of many examples of high rainfall/flood events preceding increased leptospirosis incidence (Lau et al. 2010). Massive

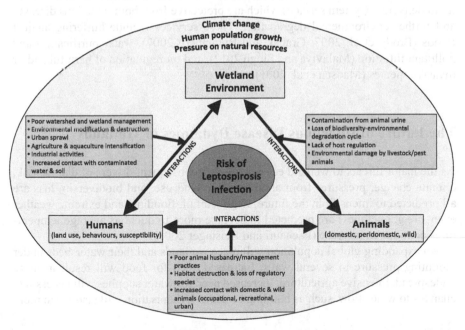

Fig. 8 Transmission ecology of leptospirosis in a wetland environment, with proximate and ultimate drivers of disease transmission. (Adapted from Lau and Jagals 2012)

flooding of urban areas brought inhabitants into contact with contaminated water, and cats, dogs and cattle were identified as the main potential reservoirs. The flood-prone situation of Jakarta (on the delta of the Ciliwung River) is itself an example of anthropogenic environmental change (high density urbanisation) of a wetland.

Increased temperatures are another form of climate change that can drive leptospirosis transmission. Warmer and more humid climates increase leptospire survival (Levett 2001; Victoriano et al. 2009; Lau et al. 2010), and may also modify human behaviours that result in increased exposure (e.g. swimming). There is also evidence that ecosystems with lower biodiversity (e.g. on remote islands, or resulting from environmental degradation), particularly those with fewer mammal species, have higher human leptospirosis incidence, potentially due to the regulation of leptospire-bearing rat populations (Derne et al. 2011).

Human population growth and the consequent pressures on natural resources, environmental change, and global climate could therefore impact on the transmission of leptospirosis via multiple pathways.

Wetland Degradation and Vulnerability: A Vicious Cycle

Altering wetland ecosystems and surrounding landscapes may drive disease risk in a twofold manner. Primary threats and disruptions are described above. However, these also compromise a wetland's resilience to environmental assaults and its ability to deliver ecosystem services which are protective from both infectious diseases and further environmental degradation. These services include buffering against floods (Brody et al. 2007; Granek and Ruttenberg 2007), water purification and pollutant filtration (Malaviya and Singh 2012), and bioregulation of harmful and/or invasive species (Madsen et al. 2001).

The Future of Infectious Disease Dynamics in Wetlands

As the major threats to wetland ecosystems and consequent drivers of disease risk, climate change, pressures from anthropogenic land use, and biodiversity loss are all predicted to intensify in the future. High rainfall, flooding, and extreme weather events (e.g. cyclones) are predicted to become more frequent as average temperatures continue to increase (Pachauri and Reisinger 2007).

An expanding global population will place wetlands and their watersheds under mounting pressure in several ways. Increased need for food will result in more widespread, intensive agriculture. Increased need for water supplies will necessitate changes to water flow, such as building of dams. Urbanisation will also mean more encroachment into wetland areas. Intensification of agriculture and urbanisation will also cause biodiversity loss. The resulting degradation of wetland environments will compromise their ability to provide disease regulation as an ecosystem service. Increased interaction between humans, other disease hosts and wetlands will also heighten exposure to wetland pathogens. Furthermore, an ever-growing need for urban, agricultural and industrial areas may outcompete conservation priorities and result in destruction and poor management of wetlands.

Conversely, the risk of water-borne diseases from wetland environments may be reduced if a deeper understanding of the links between wetland ecosystem health and human health and well-being leads to the implementation of effective management and conservation practices (Weinstein and Woodward 2005). It is recognised, however, that in real world situations, improved outcomes for both ecosystem function and human health may at times be mutually exclusive, or a wetland may be managed in a way that maximises the provision of another ecosystem service (e.g. irrigation of food crops) to the detriment of disease risk mitigation (Horwitz and Finlayson 2011). Where compromises between ecosystem services and ecosystem health of a wetland must be made, the best possible outcome can only ensue when human health consequences (including infectious disease risk) of alternative scenarios are factored into the decision making process.

Reducing Infectious Diseases in Wetlands

Some general approaches can be taken to reduce infectious disease risk from exposure to wetlands, whilst maintaining beneficial interactions with these unique environments. Firstly, measures are needed to prevent point source and non-point source contamination of wetlands by animal and human faecal matter, and by nutrient, sediment, and toxins. Adequate management of sewage and agricultural and industrial waste are therefore imperative. To minimise the runoff from agricultural pastures into wetlands, drainage systems need to consider the hydrology of the re gion—wetlands must be considered in the context of the surrounding landscape and the watercourses which connect them. Human exposure to pathogens in wetlands should first and foremost be minimised by the provision of safe drinking water, which requires robust and carefully monitored water treatment systems. Particular attention should be focused on protecting groups who are most vulnerable to water-borne infectious diseases and therefore bear the highest disease burden, e.g. populations in developing countries, particularly children.

Wetlands must be considered and valued as important ecosystems, with multiple components and linkages, which provide essential ecosystem services that regulate risk both directly and indirectly. Preservation, restoration and informed environmental management of wetlands should ensue from this underlying attitude. Ideally, environmental integrity encourages the establishment and persistence of biodiversity and a positive feedback cycle, in turn enhancing ecosystem services and human well-being.

In conjunction with these approaches, reducing risk of infectious disease in wetland environments first requires an accurate assessment and evaluation of potential risks and exposure pathways. The complex, non-linear nature of disease transmission dynamics calls for an integrated approach to environmental health impact assessments (Horwitz and Finlayson 2011; Lau and Jagals 2012). In order to minimise environmental health risks and improve health outcomes, such assessments should incorporate traditional methods of risk assessment (investigations of how environmental hazards move along exposure pathways), epidemiological assessment (identifying and quantifying links between environmental hazards and health outcomes), and health impact assessment (e.g. bio-markers of environmental exposures, or prevalence of infections or diseases) (Lau and Jagals 2012). Disease surveillance systems are thus an essential tool in not only controlling outbreaks, but also informing health impact assessment over time and space. Environmental monitoring (by using data collected on the ground or via remote sensing) is another approach; allowing the temporal and spatial relationships between specific environmental factors and human diseases to be further explored and used to identify high-risk locations and times (Carver et al. 2010).

Conclusion

This chapter has provided an overview of the diversity of infectious water-borne diseases that humans are exposed to. Infection is often a result of our intimate association with wetlands, and of the manner in which ecosystems, wider environmental and human factors all interact to determine overall risk. Every outbreak or disease emergence results from a unique set of circumstances, and understanding the drivers and how they interact to cause disease is difficult and often situation-specific. In the face of such complex systems, the need for multidisciplinary research is imperative, particularly to establish the potential health gain from interventions that are aimed at improving ecosystem health and human health concurrently. This approach follows the 'One Health' concept, which recognizes that human health is intricately linked with that of animals, and the environment in which they coexist (CDC 2013). The concept of wetlands as a setting for addressing human health determinants (Horwitz and Finlayson 2011) is particularly important in situations where ecosystem services to humans and/or preservation of ecosystem function compete. Knowledge and methods from disciplines including epidemiology, sociology, environmental health, microbiology, natural resource management, earth sciences, ecology and conservation biology are all needed to create integrated management plans for achieving the best overall outcomes of preserved wetland integrity and minimal human disease burden.

References

Aho M, Kurki M, Rautelin H et al (1989) Waterborne outbreak of Campylobacter enteritis after outdoors infantry drill in Utti, Finland. Epidemiol Infect 103:133–141

Amin MA, Fenwick A, Teesdale CH et al (1982) The assessment of a large snail control program over a 3-year period in the Gezira-irrigated-area of the Sudan. Ann Trop Med Parasitol 76:415–424

Anonymous (2013) Typhoid in Syria, March 2–8, 2013. Lancet 381:i

Appleton C, Madsen H (2012) Human schistosomiasis in wetlands in southern Africa. Wetl Ecol Manage 20:253–269

Brody SD, Highfield WE, Ryu H-C et al (2007) Examining the relationship between wetland alteration and watershed flooding in Texas and Florida. Nat Hazards 40:413–428

Cabelli VJ, Dufour AP, McCabe LJ et al (1982) Swimming-associated gastroenteritis and water-quality. Am J Epidemiol 115:606–616

Carver S, Kilpatrick AM, Kuenzi A et al (2010) Environmental monitoring to enhance comprehension and control of infectious diseases. J Environ Monitor 12:2048–2055

Carver S, Slaney DP, Leisnham PT, Weinstein P (2015) Healthy wetlands, healthy people: mosquito borne disease. In Finlayson CM et al (ed) Wetlands and human health. Springer, Dordrecht, pp 95–122

Centers for Disease Control and Prevention (CDC), Department of Health and Human Services, United States of America (2012) Parasites. Parasites—Schistosomiasis. http://www.cdc.gov/parasites/schistosomiasis/. Accessed 5 Nov 2013

Centers for Disease Control and Prevention (CDC), Department of Health and Human Services, United States of America (2013a) One health. About one health. http://www.cdc.gov/one-health/about.html. Accessed 8 Dec 2013

Centers for Disease Control and Prevention (CDC), Department of Health and Human Services, United States of America (2013b) Parasites. Parasites—Cryptosporidium (also known as 'Crypto'). http://www.cdc.gov/parasites/crypto/. Accessed 5 Nov 2013

Centers for Disease Control and Prevention (CDC), Department of Health and Human Services, United States of America (2013c) Parasites. Parasites—Giardia. http://www.cdc.gov/parasites/giardia/. Accessed 5 Nov 2013

Colwell RR (1996) Global climate and infectious disease: the cholera paradigm. Science 274:2025–2031

Conaty S, Bird P, Bell G et al (2000) Hepatitis A in New South Wales, Australia, from consumption of oysters: the first reported outbreak. Epidemiol Infect 124:121–130

Constantin de Magny G, Murtugudde R, Sapiano MR et al (2008) Environmental signatures associated with cholera epidemics. Proc Nat Acad Sci 105:17676–17681

Corvalan CF, Hales S, McMichael AAJ (2005) Ecosystems and human well-being: health synthesis. World Health Organisation. http://www.who.int/globalchange/ecosystems/ecosystems05/en/. Accessed 16 Dec 2013

Current WL, Garcia LS (1991) Cryptosporidiosis. Clinical Microbiology Reviews. 4(3):325–358

Curriero FC, Patz JA, Rose JB et al (2001) The association between extreme precipitation and waterborne disease outbreaks in the United States, 1948–1994. Am J Public Health 91:1194–1199

DeJong RJ, Morgan JA, Paraense WL et al (2001) Evolutionary relationships and biogeography of Biomphalaria (Gastropoda: Planorbidae) with implications regarding its role as host of the human bloodfluke, Schistosoma mansoni. Mol Biol Evol 18:2225–2239

Dennis DT, Smith RP, Welch JJ et al (1993) Endemic giardiasis in New Hampshire: a case-control study of environmental risks. J Infect Dis 167:1391–1395

Derne BT, Fearnley EJ, Lau CL et al (2011) Biodiversity and leptospirosis risk: a case of pathogen regulation? Med Hypotheses 77:339–344

Dufour B, Moutou F, Hattenberger AM et al (2008) Global changes: impact, management, risk approach and health measures—the case of Europe. Revue Scientifique Et Technique-Office International Des Epizooties 27:529–540

Dziuban EJ, Liang JL, Craun GF et al (2006) Surveillance for waterborne disease and outbreaks associated with recreational water-United States, 2003–2004. Centers for Disease Control and Prevention (CDC), US Department of Health and Human Services, United States of America. http://www.cdc.gov/mmwr/preview/mmwrhtml/ss5512a1.htm. Accessed 8 June 2013

Epstein PR (1993) Algal blooms in the spread and persistence of cholera. Biosystems 31:209–221

Finlayson CM, D'Cruz R (2005) Inland water systems. In: Hassan R, Scholes R, Ash N (eds) Ecosystems and human well-being: current state and trends: findings of the condition and trends working group, vol 1. Island Press, Washington, DC, pp 551–584

Fricker CR, Medema GD, Smith HV (2013) Protozoan Parasites (Cryptosporidium, Giardia, Cyclospora). World Health Organisation. http://www.who.int/water_sanitation_health/dwq/en/admicrob5.pdf. Accessed 8 August 2013

Granek EF, Ruttenberg BI (2007) Protective capacity of mangroves during tropical storms: a case study from 'Wilma' and 'Gamma' in Belize. Mar Ecol Prog Ser 343:101–105

Hansen JS, Ongerth JE (1991) Effects of time and watershed characteristics on the concentration of Cryptosporidium oocysts in river water. Appl Environ Microbiol 57:2790–2795

Horrigan L, Lawrence RS, Walker P (2002) How sustainable agriculture can address the environmental and human health harms of industrial agriculture. Environ Health Perspect 110:445–456

Horwitz P, Finlayson CM (2011) Wetlands as settings for human health: incorporating ecosystem services and health impact assessment into water resource management. Bioscience 61:678–688

Horwitz P, Finlayson M, Weinstein P (2012) Healthy wetlands, healthy people: a review of wetlands and human health interactions. Ramsar Technical Report No. 6/World Health Organization Report. Ramsar Convention Secretariat, Gland, Switzerland. http://www.ramsar.org/pdf/lib/rtr6-health.pdf. Accessed 8 June 2013

Hunter PR (2003) Climate change and waterborne and vector-borne disease. J Appl Microbiol 94:37S–46S

Johnson PTJ, Preston DL, Hoverman JT et al (2013) Biodiversity decreases disease through predictable changes in host community competence. Nature 494:230–233

Keene WE, McAnulty JM, Hoesly FC et al (1994) A swimming-associated outbreak of hemorrhagic colitis caused by Escherichia coli O157: H7 and Shigella sonnei. N Engl J Med 331:579–584

Keesing F, Belden LK, Daszak P et al (2010) Impacts of biodiversity on the emergence and transmission of infectious diseases. Nature 468:647–652

Ko WC, Cheung BMH, Tang HJ et al (2007) Melioidosis outbreak after typhoon, southern Taiwan. Emerg Infect Dis 13:896–898

Kramer MH, Sorhage FE, Goldstein ST et al (1998) First reported outbreak in the United States of cryptosporidiosis associated with a recreational lake. Clin Infect Dis 26:27–33

Kuriakose M, Eapen CK, Paul R (1997) Leptospirosis in Kolenchery, Kerala, India: Epidemiology, prevalent local serogroups and serovars and a new serovar. Eur J Epidemiol 13:691–697

Lau C, Jagals P (2012) A framework for assessing and predicting the environmental health impact of infectious diseases: a case study of leptospirosis. Rev Environ Health 27:163–174

Lau C, Weinstein P (2011) Flooding and infectious disease in rural children: Can intervention mitigate predicted increases in disease burden? Int Public Health J 2:393–404

Lau CL, Smythe LD, Craig SB et al (2010) Climate change, flooding, Urbanisation and leptospirosis: fuelling the fire? Trans R Soc Trop Med Hyg 104:631–638

Lau CL, Dobson AJ, Smythe LD et al (2012) Leptospirosis in American Samoa 2010: epidemiology, environmental drivers, and the management of emergence. Am J Trop Med Hyg 86:309–319

Levett PN (2001) Leptospirosis. Clin Microbiol Rev 14:296–326

Lillis JV, Ansdell VE, Ruben K et al (2009) Sequelae of World War II: an outbreak of chronic cutaneous nontuberculous mycobacterial infection among Satowanese islanders. Clin Infect Dis 48:1541–1546

Lipp EK, Kurz R, Vincent R et al (2001) The effects of seasonal variability and weather on microbial fecal pollution and enteric pathogens in a subtropical estuary. Estuaries 24:266–276

Lipp EK, Huq A, Colwell RR (2002) Effects of global climate on infectious disease: the cholera model. Clin Microbiol Rev 15:757–770

MacKenzie WR, Hoxie NJ, Proctor ME et al (1994) A massive outbreak in Milwaukee of Crypto-sporidium infection transmitted through the public water supply. N Engl J Med 331:161–167

Madsen H, Bloch P, Phiri H et al (2001) Bulinus nyassanus is an intermediate host for Schistosoma haematobium in Lake Malawi. Ann Trop Med Parasitol 95:353–360

Malaviya P, Singh A (2012) Constructed wetlands for management of Urban stormwater runoff. Crit Rev Environ Sci Technol 42:2153–2214

Mawdsley JL, Bardgett RD, Merry RJ et al (1995) Pathogens in livestock waste, their potential for movement through soil and environmental pollution. Appl Soil Ecol 2:1–15

Merritt RW, Walker ED, Small PL et al (2010) Ecology and transmission of Buruli ulcer disease: a systematic review. Plos Negl Trop Dis 4:e911

Meyers W, Tignokpa N, Priuli G et al (1996) Mycobacterium ulcerans infection (Buruli ulcer): first reported patients in Togo. Br J Dermatol 134:1116–1121

Mohammed HO, Wade SE (2009) The risk of zoonotic genotypes of Cryptosporidium spp. in wa-tersheds. In: Ortega-Pierres G, Caccio S, Fayer R, Mank TG, Smith HV, Thompson RCA (eds) Giardia and Cryptosporidium: from molecules to disease, vol 1. CABI, Wallingford, United Kingdom pp 123–130

Monis P, Thompson R (2003) Cryptosporidium and Giardia -zoonoses: fact or fiction? Infect Genet Evol 3:233–244

N'goran E, Diabate S, Utzinger J et al (1997) Changes in human schistosomiasis levels after the construction of two large hydroelectric dams in central Cote d'Ivoire. Bull World Health Organ 75:541

Nenonen NP, Hannoun C, Horal P et al (2008) Tracing of norovirus outbreak strains in mussels collected near sewage effluents. Appl Environ Microbiol 74:2544–2549

Nkoko DB, Giraudoux P, Plisnier P-D et al (2011) Dynamics of cholera outbreaks in great lakes region of Africa, 1978–2008. Emerg Infect Dis 17:2026–2034

Ostfeld RS, Holt RD (2004) Are predators good for your health? Evaluating evidence for top-down regulation of zoonotic disease reservoirs. Front Ecol Environ 2:13–20

Ostfeld RS, Keesing F (2000) Biodiversity and disease risk: The case of lyme disease. Conserv Biol 14:722–728

Pachauri RK, Reisinger A (2007) Climate change 2007: synthesis report. Contribution of Working Groups I, II and III to the Fourth Assessment Report of the Intergovernmental Panel on Climate Change. IPCC Geneva, Switzerland. http://www.ipcc.ch/pdf/assessment-report/ar4/syr/ar4_syr_frontmatter.pdf. Accessed 8 June 2013

Patz JA, Graczyk TK, Geller N et al (2000) Effects of environmental change on emerging parasitic diseases. Int J Parasitol 30:1395–1405

Reis RB, Ribeiro GS, Felzemburgh RDM et al (2008) Impact of environment and social gradient on leptospira infection in Urban slums. Plos Negl Trop Dis 2:e228

Rose JB, Gerba CP, Jakubowski W (1991) Survey of potable water supplies for cryptosporidium and giardia. Environ Sci Technol 25:1393–1400

Rose JB, Epstein PR, Lipp EK et al (2001) Climate variability and change in the United States: Potential impacts on water- and foodborne diseases caused by microbiologic agents. Environ Health Perspect 109:211–221

Schwartz E, Kozarsky P, Wilson M et al (2005) Schistosome infection among river rafters on Omo River, Ethiopia. J Travel Med 12:3–8

Steinmann P, Keiser J, Bos R et al (2006) Schistosomiasis and water resources development: systematic review, meta-analysis, and estimates of people at risk. Lancet Infect Dis 6:411–425

Stuart JM, Orr HJ, Warburton FG et al (2003) Risk factors for sporadic giardiasis: a case-control study in southwestern England. Emerg Infect Dis 9:229–233

Summa M, Bonsdorff C-HV, Maunula L (2012) Pet dogs, A- transmission route for human noro-viruses? J Clinl Virol 53:244–247

Sykora JL, Keleti G, Martinez AJ (1983) Occurence and pathogenicity of Naegleria Fowleri in artificially heated waters. Appl Environ Microbiol 45:974–979

Victoriano AFB, Smythe LD, Gloriani-Barzaga N et al (2009) Leptospirosis in the Asia Pacific region. BMC Infect Dis 9:147

Weinstein P, Woodward A (2005) Ecology, climate and campylobacteriosis in New Zealand. In: In: Ebi K, Smith J, and BI (eds) Integration of public health with adaptation to climate change, vol 1, 1st edn. Taylor and Francis, Leiden, pp 60–71

Weniger BG, Blaser MJ, Gedrose J et al (1983) An outbreak of waterborne giardiasis associated with heavy water runoff due to warm weather and volcanic ashfall. Am J Public Health 73:868–872

Weinstein P, Russell N, Woodward A (2000) Drinking water, ecology, and gastroenteritis in New Zealand. In: Reichard EG, Hauchman FS, Sancha AM (eds) Interdisciplinary perspectives on drinking water risk assessment and management, vol 1. International Association of Hydrological Sciences Press, Wallingford, pp 41–48

World Health Organisation (2001a) Water sanitation health. Water-related diseases—typhoid and paratyphoid enteric fevers. http://www.who.int/water_sanitation_health/diseases/typhoid/en/. Accessed 8 June 2013

World Health Organisation (2001b) Water sanitation health. Water-related diseases- hepatitis. http://www.who.int/water_sanitation_health/diseases/hepatitis/en/index.html. Accessed 8 Aug 2013

World Health Organisation (2009a) Initiative for vaccine research. Diarrhoeal diseases- campylobacter. http://www.who.int/vaccine_research/diseases/diarrhoeal/en/index2.html. Accessed 8 June 2013

World Health Organisation (2009b) Initiative for vaccine research. Diarrhoeal diseases- Enterotoxigenic Escherichia coli (ETEC). http://www.who.int/vaccine_research/diseases/diarrhoeal/en/index4.html. Accessed 8 June 2013

World Health Organisation (2009c) Quantifying environmental health impacts. Global estimates of environmental burden of disease. http://www.who.int/quantifying_ehimpacts/global/envrf2004/en/index.html. Accessed 3 May 2013

World Health Organisation (2011) World Health Organisation weekly epidemiological record. Leptospirosis: an emerging public health problem. http://www.who.int/wer/2011/wer8606.pdf. Accessed 06/08/13

World Health Organisation (2012a) Media centre fact sheets. Buruli Ulcer (Mycobacterium ulcerans infection). http://www.who.int/mediacentre/factsheets/fs115/en/. Accessed 8 June 2013

World Health Organisation (2012b) Media centre fact sheets. Cholera: fact sheet no. 107. http://www.who.int/mediacentre/factsheets/fs107/en/index.html. Accessed 5 Nov 2013

World Health Organisation (2013a) Fact file. 10 facts on cholera. http://www.who.int/features/factfiles/cholera/facts/en/index1.html. Accessed 8 June 2013

World Health Organisation (2013b) Microbial fact sheets. http://www.who.int/water_sanitation_health/dwq/GDW11rev1and2.pdf. Accessed 8 June 2013.

World Health Organisation (2013c) Media centre fact sheets. Schistosomiasis: fact sheet no. 115. http://www.who.int/mediacentre/factsheets/fs115/en/. Accessed 8 June 2013

Ecosystem Approaches to Human Exposures to Pollutants and Toxicants in Wetlands: Examples, Dilemmas and Alternatives

Pierre Horwitz and Anne Roiko

Abstract Humans can be exposed to hazardous substances in wetland ecosystems. Human health can be affected by acute or chronic exposure to toxicants, through the media of water, wetland sediments, or even air through aerosols or when sediments become desiccated and airborne, or burnt. The nature of these exposures is greatly exacerbated by human activities where pollution is involved, resulting in often complex interactions and significant challenges for environmental health practitioners. While steps can be taken to minimise the health risks resulting from such exposures, this Chapter argues that the risks can increase (sometimes dramatically) if disruption to ecosystems, and the services they provide, is profound. Two principal forms of human exposures in wetland settings are distinguished: where the type or form of exposure is determined by the service that is provided (for example when drinking water contains a pollutant), and where services are eroded, creating the conditions for exposure (for example where water purification capacities of wetlands are overwhelmed by an oversupply of nutrients, resulting in an exposure to a microbial toxin). In both cases, addressing an imbalance of ecosystem services is required to ensure any necessary interventions are effective.

Keywords Exposure · Toxicants · Risks · Hazards · Metals · Arsenic · Methylmercury · Radionuclides · Nutrients · Disinfection by-products · Cyanobacteria · Peat fires · Endocrine disrupting chemicals · Microbial toxins

Introduction

Assessments of human health risks related to environmental hazards have an understandable preoccupation with the characteristics of the hazard, exposure routes and pathways, the nature and magnitude of exposures and the characteristics of indi-

P. Horwitz (✉)
School of Natural Sciences, Edith Cowan University, 270 Joondalup Drive,
Joondalup, WA 6027, Australia
e-mail: p.horwitz@ecu.edu.au

A. Roiko
School of Medicine, Griffith University, Parklands Drive, Southport, QLD, Australia
e-mail: a.roiko@griffith.edu.au

© Springer Science+Business Media Dordrecht 2015
C. M. Finlayson et al. (eds.), *Wetlands and Human Health,* Wetlands: Ecology,
Conservation and Management 5, DOI 10.1007/978-94-017-9609-5_5

Table 1 Summary of the ways in which the exposures that people experience in wetland settings are related to the ability of wetlands to provide ecosystem services

Health risk	Relevant wetland ecosystem services	Health effects, health outcomes from ecosystem services		Disruptions to wetland-ecosystems (examples)	Examples or case studies used in this chapter
		Benefits if services are maintained or enhanced	Consequences of disruption to the services		
Ill-health due to exposure to pollution	Water purification/waste treatment or dilution Other hydrological services: hydrological maintenance of biogeochemical processes Soil, sediment and nutrient retention	Prevention of exposure to environmental contaminants Enhanced abilities to interact with wetland ecosystems to derive other benefits, like those that accrue from provisioning and cultural services, or to derive an income	Exposure to: Soil or water-borne inorganic chemicals Soil or water-borne microbial toxins Atmospheric particles or chemicals	Sewage contamination Industrial contamination Eutrophication Salinisation Acidification Depletion (drainage or resource over-extraction)	Food chain bioaccumulation (e.g. TBT, radionuclides) Acute or chronic poisoning (e.g. methylmercury) Nitrate as a human health issue Blooms of toxic Cyanobacteria Respiratory diseases from wetland/peat fires

viduals or populations that make them more vulnerable to exposure (Langley 2004). Human health can be affected by acute or chronic exposure to toxicants, through the media of water, wetland sediments, or air when sediments become desiccated and airborne, or burnt. The nature of these exposures is greatly exacerbated by human activities where pollution is involved. Wetlands are settings where ecosystem services can be provided and exposure to toxicants can be prevented, or the reverse, where exposure occurs when the services are disrupted (as described in Table 1).

The pressures on inland and coastal wetland ecosystems and their resultant degradation have considerable public health implications. In the nineteenth Century the main wetland-related health problems arose from faecal and organic pollution associated with untreated human wastewater. Today much of this contamination has been eliminated in the more developed countries. However, as described in Cook and Speldewinde (2015), improved sanitation and access to a secure supply of safe drinking water—together a highly desirable combination if the objective is to prevent exposures to water pollution and safeguard public health—still present a major challenge for a significant part of the world's population.

Despite the capacity of wetlands to purify polluted water, they do have their limits (Verhoeven et al. 2006). For example, they can only process and assimilate a certain amount of agricultural runoff, and only so much inflow from domestic and industrial wastes. As more toxic and persistent chemicals (such as PCBs, DDT and dioxins), antibiotics and other pharmaceuticals, and pesticides, some of which act as 'endocrine disrupters', are added to wetlands via sources such as untreated municipal and industrial waste, the sources of the food they supply and the water itself can be rendered unfit for consumption and pose a danger to human health. With this background three categories of toxicants are used to outline the way in which humans might be exposed to toxicants in wetland settings: (i) soil or water-borne organic and inorganic chemicals; (ii) microbial toxins; and (iii) atmospheric particles or chemicals originating from wetlands. Innumerable chemical compounds, inorganic and organic, can be important in these exposure settings; a selection of the more significant compounds and well-known contamination events are provided as examples. For each, the ways in which ecosystem services are implicated or complicated by these exposures, are explored.

Soil or Water-Borne Organic or Inorganic Chemicals

Chemical contamination of wetland ecosystems can occur anywhere along a spectrum from a rapid massive influx to a very slow almost imperceptible accrual over time, and be a result of human activities and/or other biological and geological processes. At common concentrations, most pollutants are likely to cause adverse health effects only after prolonged periods of exposure. However, there are many cases worldwide where chemical pollution of wetlands has occurred to an extent that it has been detrimental to human health either through direct ingestion of contaminated water (particularly where the wetland was a source of drinking water), or through incorporation and subsequent bioaccumulation and biomagnification of toxic chemicals within the food chain (components of which were then ingested).

Metals

There are a growing number of cases of contamination of groundwater and surface waters by metal ions from natural and anthropogenic sources. Humans can be exposed to heavy metals in a wetland setting by: (i) ingesting water, either directly from a contaminated source, or via a distribution system with corroded metallic components; (ii) exposure to dust particles; or (iii) ingesting foods where they have bioaccumulated, as described above.

Common metal ions, common sources of exposure, and their health effects are relatively well known. Most metal ions can be found in effluents from mining activities, which invariably find their way into wetlands; many also are found in human infrastructure and become associated with wetlands through leakage and drainage in urban and agricultural environments, for example:

- Aluminium, Al in water from water treatment, infrastructure, cooking appliances;
- Iron, Fe, predominantly from infrastructure and also commonly present in groundwater;
- Zinc, Zn, is used in galvanising, roofing, infrastructure;

- Copper, Cu, is used predominantly in household plumbing systems (pipes) and cooking appliances, and when corroding these become sources;
- Cadmium, Cd, is used industrially (metal plating and coating) and can also be found in sludge and fertilizers;
- Mercury, Hg, enters water from batteries, atmospheric deposition (predominantly from burning coal), gold processing, or discharge from the chlor-alkali industry;
- Lead, Pb, is found in old pipes, and also petrol and paint additive forms which are distributed by atmospheric deposition;
- Arsenic, As, most often associated with acidic groundwater; industrial uses include paints or pharmaceuticals and it is commonly found in sewage;
- Tin, Sn, organic tin compounds (Tributyltin and Triphenyltin) are toxic and are constituents of anti-fouling paints;
- Chromium, Cr, used in wood treatment, oxidising agents, corrosion inhibitors, pigments;
- Radionuclides, found for example in discharges from mining activities or fertiliser production (see Appendix).

Metals are well known for their health impacts in exposed populations (see for example Hinwood et al. 2008). Cadmium exposure has been associated with renal disease and studies have also suggested that exposure to cadmium may impact the skeleton, while lead is well known for its health impacts, such as memory deterioration, cognitive difficulties, neurological impacts and kidney damage. Concerns have been expressed by several authors on the impacts of lower levels of cadmium on bone density. Inorganic arsenic is also associated with a range of health effects including vascular disease, skin lesions at high concentrations and cancer of the bladder and kidney. Aluminium has the potential to affect the central nervous system, skeletal and haemapoietic systems of humans. Copper is an essential element for people; however, some are susceptible to the effects of increased copper exposure, such as those with Wilson's disease, renal and liver disease and infants.

Numerous examples exist of such contamination, probably none more graphic than the almost inconceivably vast scale of arsenic poisoning on the Indian subcontinent (Frisbie et al. 2002), most profoundly where water used by local populations is contaminated when naturally occurring arsenic is released to groundwater that is extracted for human consumption (see Nickson et al. 2000). Similarly in southwestern Australia humans could be exposed to heavy metals from the use of groundwater that has been affected by oxidized acid sulfate soils, to irrigate crops (Hinwood et al. 2008). Base-metal mining can result in concentrations of arsenic, cadmium, lead and zinc being raised above the levels recommended for drinking, as in the Tui and Tunakohoia streams, TeAroha, North Island, New Zealand (Sabti et al. 2000). In these cases the ecosystem service that allows for soils and sediments, and their metal and metalloids, to be retained through biogeochemical conditions rather than mobilized to the receiving environment where they become the subjects of human exposure, and supplementary water treatment, have been eroded.

Two examples of both forms of exposures and their ecosystem setting, are provided below; the cases of methylmercury (Box 1), and radionuclides (see Appendix).

Box 1: Methylmercury: Source and Effects

Mercury is rarely found in concentrations where it is problematic for human health. However, it is used in a wide range of human activities, from industrial activities, manufacturing, mining and dentistry. When it is discharged, inorganic mercury is converted to a methylated form, often in anaerobic sediments or water, and/or microbially mediated, and once in this form it is readily transported in water columns and currents and made bioavailable, since it also binds strongly to proteins. This characteristic means that it is bioaccumulated, and biomagnified in the food chain (World Health Organisation 1990). In humans the main toxic effects are the inhibition of protein synthesis, its reaction with receptor sites in the central nervous system, and profound prenatal effects on neuronal development (WHO 1990). As fish and fish products are the dominant source of methylmercury in human diets, communities that are heavily dependent on fish, in other words drawing heavily on one provisioning ecosystem service, are more vulnerable to poisoning.

An intriguing example of communities with these dependencies, in the face of methylmercury, is that summarised by Guimaraes and Mergler (2012), for the Brazilian Amazon region of the Tapajós River, where artisanal gold mining peaked in the 1980s. Researchers found high levels of mercury in fish and humans, but also in soils of the region from 'natural origins'. Dose-related deficits in motor and visual functions were found in villagers, and public health interventions began with a campaign slogan to "eat more fish that don't eat other fish", with some success. However, further testing revealed that deforestation and soil erosion caused by agricultural practices like slash and burn, mobilized mercury to rivers and lakes, meaning that the ecosystem service of soil and sediment retention was eroded; land management needed to be a central part of the intervention. In addition, an epidemiological study showed that people who ate more fruit had less mercury in their blood and hair; and the consumption of Brazil nuts with high selenium levels could also be responsible for offsetting the effects of methylmercury poisoning. Understanding the mediation effects of socio-cultural practices (including personal and political processes that influence diet), and redressing the way ecosystem services have been traded-off, become central in public health interventions.

Nutrients

Nutrients (principally nitrogen and phosphorus), in organic and inorganic forms, are arguably the chemical pollutants that have caused the most concern globally, and received the most attention (Box 2). While these nutrients occur naturally (in soil containing nitrogen-fixing bacteria, decaying plants and animal manure), one of the signatures of anthropogenic stresses is elevated nutrient levels. The most important of these sources of nutrient inputs into wetlands are runoff, erosion and leaching from fertilized agricultural areas, and sewage from domestic and industrial wastewater. Atmospheric deposition of nitrogen from combustion gases and intensive animal husbandry can also be significant (WHO 2002). An important and highly charged issue in many countries is the debate concerning farming and grazing and the degradation of waterways and groundwater aquifers (Falkenmark et al. 2007; Peden et al. 2007; Shah et al. 2007). Nitrate penetrates through soil, enters waterways and remains in groundwater for decades. This is a concern globally, as increasing stock levels, poor management practices and the clearing of riparian vegetation for further grazing allows high volumes of farm effluent, excess nutrients and chemicals to enter waterways. Eutrophication (and associated algal blooms, see below) is being reported more frequently; continental river basins in North America, Europe and Africa have elevated concentrations of organic matter (Revenga et al. 2000).

Several authors (cited in WHO 2002) have reported a complication, an interaction between chemical and microbial pollution, both involved in eutrophication, where human health consequences become exacerbated by the erosion of ecosystem services in coastal wetland ecosystems. Bacteria such as *Escherichia coli*, *Salmonella* spp or *Vibrio cholerae*, which can be harmful to human health, are found where there is organic-rich effluent, but under normal conditions, these bacteria do not survive very long in saline waters due to ecosystem services provided: limited amounts of nutrients, the exposure of bacteria to UV rays which have a bactericidal effect, and the osmolarity of sea water which is much higher than that of bacteria. Algal blooms, a product of eutrophication, can erode these services: food becomes abundant, light is diminished reducing UV radiation in the water column and it has been recently reported that some algae may even release chemicals that produce osmo-protection for the bacteria. The new conditions can allow bacteria to survive and perhaps multiply, increasing the risks of human exposure.

Box 2: Nutrient enrichment in inland waters (sourced from Millennium Ecosystem Assessment, 2005 unless otherwise stated)

Over the past four decades, excessive nutrient loading has emerged as one of the most important direct drivers of ecosystem change in terrestrial, freshwater, and marine ecosystems. While the introduction of nutrients into ecosystems can have both beneficial and adverse effects, the beneficial effects will eventually reach a plateau as more nutrients are added (that is, additional inputs will not lead to further increases in crop yield), while the harmful effects will continue to grow.

Increase in nitrogen fluxes in rivers to coastal waters due to human activities relative to fluxes prior to the industrial and agricultural revolutions have been shown for many areas (Howarth and Ramakrishna 2005). Synthetic production of nitrogen fertilizer has been an important driver for the remarkable increase in food production that has occurred during the past 50 years. World consumption of nitrogenous fertilizers grew nearly eightfold between 1960 and 2003. As much as 50 % of the nitrogen fertilizer applied may be lost to the environment, depending on how well the application is managed.

Phosphorus application has increased threefold since 1960, with a steady increase until 1990 followed by a leveling off at a level approximately equal to applications in the 1980s.

Since excessive nutrient loading is largely the result of applying more nutrients than crops can use, it harms both farm incomes and the environment.

Many ecosystem services are reduced when inland waters and coastal ecosystems become eutrophic. Water from lakes that experience algal blooms is more expensive to purify for drinking or other industrial uses. Eutrophication can reduce or eliminate fish populations. Possibly the most apparent loss in services is the loss of many of the cultural services provided by lakes. Foul odours of rotting algae, slime-covered lakes, and toxic chemicals produced by some blue-green algae during blooms keep people from swimming, boating, and otherwise enjoying the aesthetic value of lakes.

Increased nitrogen fluxes (Box 2) are partly due to a dramatic and rapid global increase in nitrate fertilizer application, as well as indirect sources of nitrate (where organically bound N can be mineralised by soil bacteria into ammonia (slow), then nitrification to nitrate (rapid)(Gray 2008). Nitrate can leach into surface or groundwater, and nitrate pollution of groundwater is getting worse in northern China, India and Europe (Revenga et al. 2000; Vorosmarty et al. 2005; Shah et al. 2007). Nitrate is an important consideration for human health in three ways: (i) as a contributor to eutrophication and the problematic consequences due to prolific algal growth (see above); (ii) as a common component of food nitrate itself is relatively harmless, however nitrate can be reduced to nitrite (either by the acidic conditions found in the stomach, or by commensal bacteria in the saliva, small intestine and colon). Acidic production of N-nitroso-compounds might be problematic for human health: while these compounds are known to be carcinogenic in multiple organs in more than 40 species including higher primates (Jakszyn et al. 2006), epidemiological data supporting a causal relationship between nitrates and cancer in humans is generally lacking (Jakszyn et al. 2006; Gray 2008); and (iii) when nitrite combines with haemoglobin, debilitating its oxygen carrying function. Methaemoglobinaemia is the syndrome associated with acute expressions, and it can be fatal, particularly for infants < 3 months old (who are especially susceptible due to different respiratory pigments; Manassaram et al. 2007). Cofactors like the presence of microbial contamination, diarrhoea or respiratory diseases, may be implicated in acute cases (Gray 2008).

The Case of Disinfection By-Products (DBP's)

In the 1970s the US Environmental Protection Agency found hundreds of organic chemicals in drinking water sources, many of which were believed to be carcinogenic and teratogenic (Okun 1996). Epidemiological studies in New Orleans at this time revealed higher levels of cancer in individuals using the treated water supply versus those using untreated groundwater (Talbot and Harris 1974). This led to the passage of the Safe Drinking Water Act in the US in 1974. At the same time on the other side of the world, Rook (1974) showed that the common chemical used for disinfection in water treatment—chlorine—created disinfection by-products (DBPs) which were carcinogenic in rodents. Decades of further research summarised in Hrudey (2009) and Richardson and Postigo (2012) has revealed that over 600 DBPs (including trihalomethanes, haloacetic acids, bromide and chlorite) have been reported in drinking water and are an unintended consequence of using chemicals such as chlorine, chloramines, ozone and chlorine dioxide (and their combinations) to kill harmful pathogens in water intended for potable uses and that DBPs are formed when these chemical disinfectants react with organic matter, bromide and iodide found in the source water. Humans are exposed not only through ingesting drinking water, but also through inhalation and dermal routes while showering and swimming. To date, epidemiological data indicate potential developmental, reproductive, or carcinogenic health effects such as bladder cancer, miscarriage and birth defects in humans exposed to DBPs (Malcolm et al. 1999; Anderson et al. 2002). These data remain inconclusive with evidence of a causal relationship and support from toxicological studies lacking (Hrudey 2009).

This highlights a significant dilemma for risk management: whether to expose the population to the inevitable health risks (of varying severity) associated with undisinfected drinking water due to disease causing microbes, or face the severe but uncertain health risks associated with DBPs (Hrudey 2009). A third option avoids this dilemma; an ecosystem approach focuses on the continual and improved attention to the management of catchments. In doing so the aims would be to (i) maintain the ecological functions and processes in the catchment in order to be able to assimilate nutrients and regulate microbial pathogens and thereby minimize the need for disinfectants, and (ii) minimise organic pollution in source waters thereby reducing the reactive byproducts if and when disinfectants need to be used. This reasoning draws on several ecosystem services (water purification, disease control and regulation, soil, sediment and nutrient retention) and presents a prudent, precautionary approach to reducing health risks from DBPs.

Endocrine disrupting chemicals (EDCs) in aquatic ecosystems, particularly those ecosystems used for potable water presents another area of controversy of exposure assessments. EDCs are chemicals that can mimic and/or interfere with the action of natural hormones in organisms. Pollutants that contain EDCs include pesticides, dioxins, excreted pharmaceuticals, alkylphenols, furans and some metals, which enter the environment directly via agricultural or industrial activities or from treated sewage. Heightened levels of concern and research into EDC's have triggered by the push for potable reuse of treated wastewater. Focussing on estrogenic EDC's, Leusch et al. (2009) found that while EDC's are present in treated waste water in low concentrations, they are also ubiquitous in air, soil and food and even cosmetics and medical applications and that exposure to humans through drinking recycled water is likely to be insignificant compared to current dietary intakes. Nevertheless, ongoing research is needed to assess and manage potential health risks to both aquatic organisms and humans (Melnick et al. 2002; Leusch et al. 2009).

Soil or Water-Borne Microbial Toxins

Some forms of pollution emanate from the metabolic by-products or breakdown products of microbes, and particularly ecosystems suffering anthropogenic stress. Probably the most relevant examples are toxins associated with blooms of Cyanobacteria (sometimes called "blue green algae"), which occur in freshwater, estuarine, and near-shore marine wetland ecosystems. Harmful algal blooms, attributed partly to nutrient loads, have increased in freshwater and coastal systems over the last 20 years (UNEP 2007). Toxigenic Cyanobacteria (those species that have toxic strains or populations) are capable of producing neurotoxins (acting specifically on nerve cells of vertebrates), hepatotoxins (damage the metabolic processes in the liver), dermatotoxins (skin irritants) and endotoxins (gastrointestinal irritants) (Carmichael 2002). In addition to the production of toxins, Cyanobacteria have often been associated with the production of taste and odour compounds such as geosmin and 2-methylisoberneol (2-MIB), particularly where drinking water is sourced direct from a wetland ecosystem (Hrudey and Hrudey, 2007). Exposure to cyanobacterial toxins through consumption of contaminated drinking water has however also resulted in poisoning:

> The earliest demonstration of this was in 1983, when the population of a rural town in Australia was supplied with drinking water from a reservoir carrying a dense water bloom of a toxic species of cyanobacterium, *Microcystis aeruginosa*. The toxicity of this water bloom was being monitored in the reservoir. The controlling authority dosed the reservoir with copper sulphate to destroy the cyanobacteria, which caused the cells to lyse and release toxin into the water. Epidemiological data for liver injury in the affected population, a control population and comparison of the time periods before the bloom, during the bloom and lysis, and afterwards, showed clearly that liver damage had occurred only in the exposed population and only at the time of the water bloom. (quote from Falconer and Humpage 2005)

Fristachi et al. (2008) suggest that these more obvious, often acute, impacts on human health need to be supplemented with a consideration of less well-known chronic, subtle or insidious impacts, and potential impacts where hazards exist in remote areas where health impacts have not yet been sustained. Most commentators on the occurrence of cyanobacterial blooms suggest that nutrient enrichment is an important causal agent, but beyond that many other biophysical parameters are involved, including temperature, light availability, meteorological conditions, alteration of water flow, turbidity, vertical mixing, pH changes and trace metals, such as copper, iron and zinc (Fristachi et al. 2008).

Atmospheric Particles or Chemicals

Hydrological change to wetland ecosystems that results in aerosol production has been demonstrated to have impacts on human health. A graphic example of this is the Indonesian peat fires of 1997 (see Box 3). From these fires a significant number of cases of asthma, bronchitis and acute respiratory infection were reported

(Kunii et al. 2002). Sensitive sub-groups such as the elderly and those with pre-existing illness reported a greater severity in symptoms (Kunii et al. 2002). In Singapore, impacts from the Indonesian fires were also observed with increases in outpatient attendance for respiratory illnesses including asthma (Emmanuel 2000). Mott et al. (2005) investigated re-admission of older patients associated with smoke exposure from the 1997 fires in South East Asia specifically Malaysia. They reported short-term increases in re-admission of patients with cardio-respiratory and respiratory diseases. Frankenberg et al. (2005), using data from an Indonesian population-based longitudinal survey combined with satellite measures of aerosol levels, assessed the impact of smoke from the fires on human health. Their results indicated that exposure to the smoke from fires has a negative and significant (but mostly transitory) impact on the health of older adults and prime age women.

Box 3: Peat fires in Indonesia (sourced and adapted from Rieley 2004)

Tropical peatlands are one of the largest carbon stores on earth, the release of which has implications for climate change (Page et al. 2002). The vast majority of these peatlands are lowland, rain-fed ecosystems with a natural vegetation cover of peat swamp forest (PSF). In a natural state, lowland tropical peatlands support a luxuriant growth of rainforest trees up to 40 m tall, overlying pet deposits up to 20 m thick, but any persistent environmental change, particularly decrease in wetness, threatens their stability and makes them susceptible to fire. At the present time, the peatlands of Southeast Asia represent a globally important carbon store, which has accumulated over 26,000 years or more. In recent decades, however, an increasing proportion of this store has been converted to a carbon source, through a combination of deforestation, land-use change and fire.

Fires were widespread on the extensive peatlands of Indonesia during the 1997 El Niño and recurred in 2002, 2004 and 2006. By using satellite imagery and ground measurements within a 2.5 million ha study area in Central Kalimantan it was determined that 32 % (0.79 M ha) of the area burned in 1997, of which peatland accounted for 91.5 % (0.73 M ha), releasing 0.19–0.23 Gt of carbon to the atmosphere through peat combustion. It was estimated that between 0.81–2.57 Gt of carbon were released to the atmosphere from Indonesia's peatlands in 1997 as a result of burning peat and vegetation.

Many of these fires spread into forest areas where they burned with great intensity. In Kalimantan, South Sumatra, and West Papua, fires were started on, or reached areas of peatland, burning both the vegetation and the underlying peat. In Central Kalimantan, the situation was exacerbated by a massive peatland conversion project—the so-called Mega Rice Project (MRP). This scheme was initiated in 1995 with the aim of converting 1 M ha of wetland, mostly peatland, to agricultural use. Throughout the MRP area extensive, deep drainage and irrigation canals were excavated and much of the peat swamp forest was logged and, during 1997 fire was being used as a rapid land clearance tool. Initial estimates indicated that approximately 4.5 M ha of land had been damaged by the 1997 fires, but more detailed assessments doubled this figure to 9 M ha. Of this latter area, as much as 1.45 M ha was believed to be peat and swamp forest although no one made credible estimates of the area of peatland affected by fire at the time.

The two most intensive sources of smoke and particulate matter were the fires centred on the peatlands of Central Kalimantan and the Riau area of South Sumatra. Here both vegetation and underlying peat caught fire, contributing greatly to the so-called haze (particulate-laden smog), which blew north-westwards to affect Singapore and Malaysia. During this time solar radiation in Central Kalimantan was reduced to 40 % of normal levels whilst visibility was reduced to 25 m.

It has been estimated that the financial consequences of the fires were over US$ 3 billion from losses in timber, agriculture, non-timber forest products, hydrological and soil conservation services, and biodiversity benefits, whilst the haze cost an additional US$ 1.4 billion, most of which was borne by Indonesians for health treatment and lost tourism revenues.

The widespread peatland fires that occurred throughout Indonesia during the strong ENSO-related drought of 1997 resulted in the combustion of stored carbon that took between 1000 and 2000 years to accumulate (Page et al. 2002). At the current estimated rate of carbon accumulation in Central Kalimantan peatlands of 85 g m^{-2} year^{-1}, this single fire event represents an approximate loss of between 70–200 years of carbon sink function. The Southeast Asian region is currently subject to increasing climatic variability and seasonal precipitation extremes associated with future ENSO-events are predicted to become more pronounced (Goldammer and Price 1998; Siegert et al. 2001). This may lead to reduced water supply to and retention by peatlands, leading to a lowering of water tables. This will limit the rate of peat accumulation where it is still taking place, enhance degradation and oxidation on peatlands that are no longer actively forming peat, and greatly increase the likelihood of peatland fires, with consequent rapid loss of stored carbon. Increased climatic seasonality and variability has the potential to switch the tropical peatland ecosystems of South-east Asia from carbon sinks to carbon sources.

Unless land use policies are changed to control logging and the drainage and clearing of peatland for plantations, recurrent fires will lead to a loss of Indonesia's peat swamp forests and continued, high emissions of CO_2 to the atmosphere.

The information given above has been derived from Rieley (2004).

Conclusions

This Chapter presents examples of pollutants and other toxicants that commonly occur in wetland settings, and where human exposure to them can result in negative health outcomes. Two principal forms of human exposures in wetland settings are distinguished: where the type or form of exposure is determined by the service that is provided (for example when drinking water contains a contaminant), and where services are eroded creating the conditions for exposure (for example where water purification capacities of wetlands are overwhelmed by an oversupply of nutrients, resulting in an exposure to a microbial toxin). In both cases, understanding the mediating effects of socio-cultural practices, and addressing an imbalance of eco-system services, is required to ensure any necessary interventions to reduce human exposures are effective.

A risk assessment/risk management paradigm of managing health risks from environmental chemicals would traditionally be based on how much chemical exposure humans and ecosystems can absorb without damage. An ecosystem approach, which determines the upstream socio-cultural and political causes of the toxicant accumulation and exposures, and examines the wetland setting and the trade-offs that are made between ecosystem services, offers an alternative to this paradigm. Restoring ecosystems and their attendant ecosystem services, would address many of the upstream factors responsible for hazardous environmental exposures to pollutants and other toxicants by humans. Such restoration works are rarely considered within the scope of public health interventions, yet should be.

Appendix

Case Study: Assessment of the Potential Radiological Exposure of Humans Downstream of an Operating Uranium Mine in Australia.

Andreas Bollhöfer and Che Doering, Environmental Research Institute of the Supervising Scientist (eriss), Darwin, Australia.

Context Kakadu National Park is part of the Alligator Rivers Region (ARR) in the Northern Territory of Australia (Fig. 1) and contains a diverse range of wetland types, from intertidal forested wetlands and mudflats, to seasonal freshwater marshes, rivers and permanent billabongs (Ramsar Convention 2012). The wetlands of Kakadu are important for the local Aboriginal people for cultural and recreational uses and have been used for thousands of years for customary harvesting of bush foods. Research conducted in the late 1970s (Altman 1987) identified 90 animal and 80 plant species, including water buffalo, fish and shellfish, that were regularly harvested as bush foods, within the broader regional context, many of them associated with the large wetlands. Thirty years later, customary harvesting in the region had changed little and was still of crucial importance to the local economy (Altman 2003).

Kakadu National Park was declared in several stages from 1979 following recommendations of the Ranger Uranium Environmental Inquiry (RUEI 1977) to grant Aboriginal land rights and to establish a national park over the same area where development of a uranium mine had been proposed. At the same time the statutory

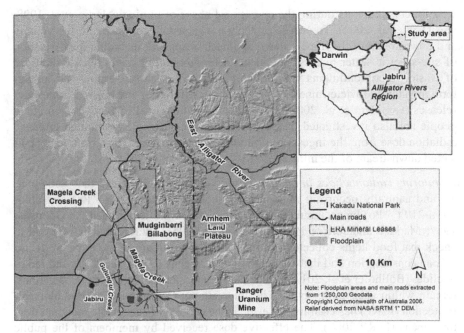

Fig. 1 The study area in Northern Australia

Office of the Supervising Scientist for the ARR was established under the Environment Protection (Alligator Rivers Region) Act 1978 to provide assurance that the environment of the ARR (including Kakadu) was protected from the potential impacts of uranium mining. Kakadu has since been designated a Wetland of International Importance under the Ramsar Convention and inscribed on the World Heritage List of the United Nations Educational, Scientific and Cultural Organization (UNESCO) for its natural and cultural values.

With the development of Ranger mine, a regional centre, Jabiru, was established as a base for workers and associated services, and developed as a hub for the influx of tourists visiting Kakadu. Many Aboriginal people moved to Jabiru but there are various outstations and smaller communities scattered throughout the region. The Kakadu National Park Management Plan supports hunting and gathering of wild food items by Aboriginal people living in and around the region if it takes place in accordance with long-standing tradition (Director of National Parks 2007). Although some market food has replaced traditional staple foods, bush foods sourced from the wetlands of Kakadu are still an important component of the Aboriginal diet in the region.

The location in the wet-dry tropics means that water management is a major issue at Ranger mine. The release of water from the site into the downstream environment has been minimised by the use of a series of retention ponds and other water treatment technologies. Magela Creek borders the site and is a receiving waterway for potential releases of waste water from the mine. The creek has been the focus of various monitoring programs due to the importance of the aquatic transport pathway for contaminants during the operational phase of mining (Martin et al. 1998; van Dam et al. 2002). To ensure protection of aquatic ecosystems in the vicinity and downstream of the mine a tiered approach is applied, including the derivation of site-specific water quality guidelines, determination of 'safe' release dilutions of waste water, continuous water quality monitoring, early warning and longer-term monitoring to determine the ecological significance of any impacts from water releases (van Dam et al. 2002). The potential radiological impacts to Aboriginal people are also investigated, based on an assessment of the committed effective radiation dose from the ingestion of radionuclides accumulated in bush foods collected downstream of the mine site.

Monitoring radionuclides in aquatic bush foods It is well known that some metals and naturally occurring radionuclides such as radium-226 (^{226}Ra) and polonium-210 (^{210}Po) bioaccumulate in aquatic biota, in particular shellfish (Johnston 1987; Markich and Jeffree 1994; Jeffree et al. 1995). Therefore, it is essential to check that food items collected from wetlands downstream of the mine are fit for human consumption and that concentrations of metals (Bollhöfer 2012) and radionuclides (Bollhöfer et al. 2011) in organism tissues, attributable to mine operations, remain within acceptable levels. The International Commission on Radiological-Protection (ICRP) has established levels in the context of acceptable radiation dose to humans (ICRP 2007). The effective dose received by members of the public who are incidentally exposed to radiation from practices such as uranium mining should not exceed 1 mSv in a year above the natural background level. Additionally, to account for situations where a person may be incidentally exposed to radiation from multiple practices, a dose constraint of 0.3 mSv for a single practice is recommended. Enhanced body burdens of mine-derived metals and radionuclides in biota may potentially also reach levels that harm the organisms themselves, and any elevation in tissue concentrations can provide early warning of bioavailability of these constituents. Hence, a bioaccumulation monitoring program also serves an ecosystem protection role in addition to the human health aspect. Specific metrics are used; for instance, radionuclide transfer from the environment is typically quantified as the concentration ratio (CR) between the radionuclide activity concentration in biota tissue (or the whole organism) and that in soil (for terrestrial organisms) or water (for aquatic organisms) that often function as reservoirs for nutrients and contaminants.

Mudginberri Billabong is the first permanent water body 12 km downstream of the mine along Magela Creek. The billabong is accessed by Aboriginal people to harvest freshwater mussels (*Velesunio angasi*) and fish (various species) and determining the levels of radionuclides and metals in these food items has been part of the Supervising Scientist's monitoring program. The focus has in particular been directed at mussel tissue analysis for ^{226}Ra, as it has been shown that ^{226}Ra in freshwater mussels is the biggest potential contributor to mine-related ingestion dose from a hypothetical release of retention pond waters from the mine site (Martin 2000; van Dam et al. 2002). This is primarily because (a) freshwater mussels are an integral part of the diet of local Aboriginal people, (b) the high uptake factor, or

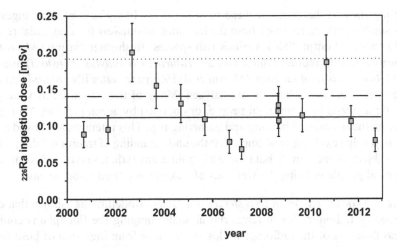

Fig. 2 Annual committed effective dose from ingestion of ^{226}Ra for a 10 year old child eating 2 kg of mussels per year. The median for all data (*solid line*), the 80th percentile (*dashed*) and the 95th percentile (*dotted*) are shown as well

Cr, of 19,000 (Johnston 1987) for radium in freshwater mussels and (c) the large ingestion dose coefficient for ^{226}Ra of 0.8 μSv · Bq^{-1} (for a 10 year old child) (ICRP 1995). Mussels are collected annually from the billabong and then aged, processed and analysed for their radionuclide content. In Fig. 2, the estimated annual committed effective radiation dose from the consumption of ^{226}Ra in 2 kg of mussels is shown.

It is important to emphasize that Fig. 2 shows the total dose from ingestion of ^{226}Ra in mussels, rather than a dose related to mining operations, which is assessed by routinely monitoring the ^{226}Ra activity concentration in Magela Creek waters upstream and downstream of the mine. A ^{226}Ra activity concentration limit for Magela Creek water of 10 mBq · L^{-1} above typical background has been set based on potential dietary uptake of ^{226}Ra by Aboriginal people downstream of the mine (Sauerland et al. 2005): if 2 kg of mussel flesh were ingested by a 10 year old child in a year and the ^{226}Ra activity concentration in Magela Creek water was 10 mBq·L^{-1} above background, then the ingestion of ^{226}Ra in freshwater mussels would lead to an additional committed effective dose of 0.3 mSv. To this end however no increase of ^{226}Ra activity concentrations in Magela Creek water downstream of the mine has been observed and mine related dose from the ingestion of radionuclides in mussels is negligible (Supervising Scientist 2012). Although the ^{226}Ra concentrations in mussels from Mudginberri Billabong are higher than in a control billabong nearby, this is associated with the low water Ca and Mg concentrations in Magela Creek (Bollhöfer et al 2011).

Many other aquatic biota have been analysed for radionuclides over the past three decades, as they have been identified as traditional food items for Aboriginal people in the region and people worried about whether they were safe to eat. Although modeling showed that none of these food items are likely to contribute ingestion doses similar to those received from the ingestion of mussels for a potential release of mine water (Martin 2000), various fish species, freshwater shrimp (*Macrobrachiumrosenbergii*), reptiles (*Acrochordus arafurae, Crocodylus johnstoni*), magpie goose (*Anseranas semipalmata*) (Martin et al. 1998) and water lily (*Nymphaea spp*) (Pettersson et al. 1993) have been analysed. Many of these food items were provided for analysis by Aboriginal people, or obtained by accompanying Aboriginal people on the occasional hunting and gathering trip. This interaction allowed a two way knowledge exchange and improved the understanding of traditional diet, which is still subject to uncertainty but ultimately guides any radiation dose assessment for Aboriginal people utilising the wetlands of Kakadu as a food resource base.

Monitoring radionuclides in terrestrial bush foods Modelling has shown that consumption of radionuclides in terrestrial animals foraging the floodplains contribute less than 1 % of the radiological dose to humans from ingestion of bush foods after a hypothetical release of mine waters and is negligible for current operation of the mine (Martin 2000). However, recent emphasis has been on investigating the uptake of radionuclides in terrestrial plant and animal species due to the increased radiological importance that these bush foods are expected to have during the post-rehabilitation phase when people access the rehabilitated mine site and a terrestrial landform re-establishes. It has also been shown that radionuclides and metals released from Ranger into Magela Creek do not deposit in the sandy creek beds between Ranger mine and Mudginberri Billabong, but ultimately on the Magela Floodplain with an area of approximately 220 km^2 (Murray et al. 1993). This deposited activity is available for uptake from the floodplain sediment and soils by terrestrial animals and plants, when the floodplains dry during the dry season.

Samples from terrestrial animals and plants continue to be investigated to provide re-assurance to Aboriginal people and investigate radionuclide transfer characteristics. The magnitudes of CRs influence the importance of a bushfood-radionuclide combination for ingestion dose assessment and Fig. 3 shows a summary of CRs for fruit (figs, plums, passion fruit, apples) and yams in the ARR.

Yam CRs are similar for all radionuclides and generally greater than the corresponding CRs for fruit (see also Ryan et al. 2005). Fruit CRs vary over two orders of magnitude and followed the approximate order ^{226}Ra ≈ lead-210 (^{210}Pb) > ^{210}Po > thorium-232 (^{232}Th) > uranium-238 (^{238}U). Some elements when absorbed from the soil through the root system are characterised by low mobility within the plant and are retained and accumulated in the roots (Pöschl and Nollet, 2007). Translocation of uranium, thorium, radium, lead and polonium from the roots to the fruit is limited (Mitchell et al. 2013), which explains the higher CRs for yam compared to the fruits.

Wallaby (*Macropus agilis*), pig (*Sus scrofa*) and feral buffalo (*Bubalusbubalis*) tissue have also been analysed including some organ tissues, as organs are routinely consumed by Aboriginal people (Altman 1987). CRs for ^{210}Po and ^{210}Pb in organs are typically higher than for flesh by approximately one order of magnitude. The liver and kidneys act as filters for substances penetrating or leaving the body of an

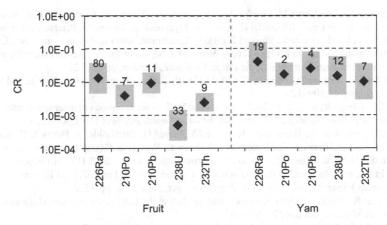

Fig. 3 Concentration ratio (*CR*) geometric means (*black diamonds*) and standard deviations (*grey columns*) for fruits and yams on a fresh weight bush food to dry weight soil basis (*numbers* indicate the number of bush food samples on which the geometric mean is based) (from Doering and Bollhöfer 2014)

animal and have been shown to accumulate various pollutants. In contrast ^{226}Ra preferentially accumulates in the bones (Pöschl and Nollet 2007). Analysis of the data indicates that ^{210}Po is the most important radionuclide in terrestrial animals for human ingestion dose assessment, due to its relatively high CR in terrestrial bush foods and high ingestion dose coefficient.

The monitoring of radionuclides in bush foods sourced from the wetlands of Kakadu National Park has been used to assess radiological impacts to Aboriginal people from a traditional diet. The results demonstrate that there are no unacceptable radiological impacts and provide assurance to Aboriginal people that bush foods in Kakadu are safe to eat from a radiological perspective.

References

Anderson DM, Overpeck JT, Gupta AK (2002) Increase in the Asian SW monsoon during the past four centuries. Science 297(5581):596–599

Carmichael WW (2002) Health effects of toxin producing Cyanobacteria: the "Cyanohabs". In: Workshop. Freshwater harmful algal blooms: health risk and control management. Instituto Superiore di Sanità. Rome, 17 October 2000. (Proceedings edited by S. Melchiorre, E. Viaggiu and M. Bruno), p103

Cook A and Speldewinde P (2015). Public health perspectives on water systems and ecology. Chapter 2 In: Finlayson M, Horwitz P, and Weinstein P (eds) Wetlands and Human Health. Springer, New York.

Emmanuel SC (2000) Impact to lung health of haze from forest fires: the Singapore experience. Respirology 5(2):175–182

Falconer IR, Humpage AR (2005) Health risk assessment of cyanobacterial (blue-green algal) toxins in drinking water. Int J Environ Res Public Health 2(1):43–50

Falkenmark M, Finlayson CM, Gordon LJ, Bennett EM, Chiuta TM, Coates D, Gosh N, Gopal-akrishnan M, de Groot RS, Jacks G, Kendy E, Oyebande L, Moore M, Peterson JM, Portugez GD, Seesink JM, Thame K, Wasson R (2007) Agriculture, water and ecosystems: avoiding the costs of going too far. In Molden D (ed) Water for food, water for life: a comprehensive assessment of water management in agriculture. Earthscan, London, pp 233–277

Frankenberg E, McKee D, Thomas D (2005) Health consequences of forest fires in Indonesia. Demography 42:109–129

Frisbie SH, Ortega R, Maynard DM, Sarkar B (2002) The concentrations of arsenic and other toxic elements in bangladesh's drinking water. Environ Health Perspect 110:1147–1153

Fristachi A, Sinclair JL, Hambrook-Berkman JA, Boyer G, Burkholder J, Burns J, Carmichael W, du Four A, Frazier W, Morton SL, O'Brien E, Walker S (2008) Occurrence of Cyano-bacterial Harmful Algal Blooms working group report. In: Hudnell HK (ed) Proceedings of the Interagency International Symposium on Cyanobacterial Harmful Algal Blooms, vol 619. Advances Experimental Medical Biology) Springer, New York, pp 37–97

Goldammer JG, Price C (1998) Sea level changes during the late Pleistocene and Holocene in the Strait of Malacca. Nature 278:441–443

Gray NF (2008) Drinking water quality, 2nd edn. Cambridge University Press, Cambridge

Guimaraes JRD, Mergler D (2012) A virtuous cycle in the Amazon: reducing mercury exposure from fish consumption requires sustainable agriculture. In: Charron D (ed) Ecohealth in re-search and practice: innovative applications of an ecosystem approach to health. Springer, International Development Research Centre, Ottawa

Hinwood A, Horwitz P, Rogan R (2008) Human exposure to metals in groundwater affected by acid sulphate soil disturbance. Arch Environ Contam Toxicol 55:538–545

Howarth R, Ramakrishna K (contributing lead authors) (2005) Nutrient Management (Chapter 9). In Ecosystems and Human Well-Being: Policy Responses (Vol. 3-Findings of the Responses Working Group). Millennium Ecosystem Assessment: Island Press

Hrudey SE (2009) Chlorination disinfection by-products, public health risk, tradeoffs and me. Water Res 4 3:2057–2092

Hrudey SE, Hrudey EJ (2007) A nose for trouble—the role of off-flavours in assuring safe drinking water. Water Sci Technol 55(5):239–247

Jakzsyn P, Bingham S, Pera G, Agudo A et al (2006) Endogenous versus exogenous exposure to N-nitroso compounds and gastric cancer risk in the European prospective investigation into cancer and nutrition (EPIC-EURGAST) study. Carcinogenesis 27(7):1497–1501

Kunii O, Kanagawa S, Yajima I, Hisamatsu Y, Yamamura S, Amagai T, Ismail IT (2002) The 1997 haze disaster in Indonesia: its air quality and health effects. Arch Environ Health 57(1):16–22

Langley A (2004) Risk assessment. In: Cromar N, Cameron S, Fallowfield H (eds) Environmental health in Australia and New Zealand. Oxford University Press, Melbourne, pp 92–110

Leusch F, Moore MR, Chapmann H (2009) Balancing the budget of environmental estrogen expo-sure –the contribution of recycled water. Water Sci Technol—(WST) 60(4):1003–1012

Malcolm MS, Weinstein P, Woodward AJ (1999) Something in the water? A health impact assess-ment of disinfection by-products in New Zealand. NZ Med J 12:404–407

Manassaram DM, Backer LC, Moll DM (2007) A review of nitrates in drinking water: maternal exposure and adverse reproductive and developmental outcomes. Ciên Saúde Colet 12(1): 153–163

Melnick R, Lucier G, Wolfe M, Hall R, Stancel G, Prins G, Gallo M, Reuhl K, Ho S-M, Brown T, Moore J, Leakey J, Haseman J, Kohn M (2002) Summary of the national toxicology program's report of the endocrine disruptors low-dose peer review. Environ Health Perspect 110(4): 427–431

Millennium Ecosystem Assessment (2003) Ecosystems and human well-being: a framework for assessment. Island Press, Washington, DC

Millennium Ecosystem Assessment (2005) Millennium ecosystem assessment synthesis report. Island Press, Washington, DC

Mott JA, Mannino DM, Alverson CJ, Kiyu A, Hashim J, Lee T, Falter K, Redd SC (2005) Cardio-respiratory hospitalizations associated with smoke exposure during the 1997, Southeast Asian forest fires. Int J Hyg Environ Health 208(1–2):75–85

Nickson RT, McArthura JT, Ravenscroft P, Burgess WG, Ahmed KM (2000) Mechanism of arsenic release to groundwater, Bangladesh and West Bengal. Appl Geochem 15:403–413

Okun DD (1996) From cholera to cancer to cryptospiridiosis. J Environ Eng 122:453–458

Page SE, Siegert F, Rieley JO, Boehm HD-V, Jaya A, Limin S (2002) The amount of carbon released from peat and forest fires in Indonesia during 1997. Nature 420:61–65

Peden D (2007) Water and livestock for human development. In Molden D (ed) A comprehensive assessment of water management in agriculture. Earthscan, Washington, DC, pp 485–514

Revenga C, Brunner J, Henninger N, Kassem K, Payne R (2000) Pilot analysis of global ecosystems: freshwater ecosystems. World Resources Institute, Washington, DC

Richardson SD, Postigo C (2012) Drinking water disinfection by-products. Emerging organic contaminants and human health—the handbook of environmental chemistry, pp 93–137

Rieley J (2004) Peat fires in Indonesia: the facts. http://www.imcg.net/media/newsletter/nl0404. pdf. Accessed 19 Jan 2015

Rook JJ (1974) Formation of haloforms during chlorination of natural waters. Water Treat Exam 23:234–243

Sabti H, Hossain MM, Brooks RR, Stewart RB (2000) The current environmental impact of base-metal mining at the Tui Mine, TeAroha, NewZealand. J R Soc N Z 30:197–208

Shah T, Burke J, Vilholth K (2007). Groundwater: a global assessment of scale and significance. In: Molden D (ed) A comprehensive assessment of water management in agriculture Earthscanand Colombo: International Water Management Institute, London, Colombo, pp 395–423

Siegert F, Ruecker G, Hindrichs E, Hoffman AA (2001) Increased damage from fires in logged forests during droughts caused by El Niño. Nature 414:437–440

Talbot P, Harris RH (1974) The implications of cancer-causing substances in Mississippi river water. (A Report by the Environmental Defense Fund). Environmental Defense Fund, Washington, DC

United Nations Environment Program UNEP (2007). Global environment outlook 4 environment for development, UNEP, Nairobi

Verhoeven JTA, Arheimer B, Yin C, Hefting MM (2006) Regional and global concerns over wetlands and water quality. Trends Ecol Evol 21:96–103

Vörösmarty CJ, Lévêque C, Revenga C (2005) Fresh water. In: Hassan R, Scholes R, Ash N (eds) Ecosystems and human well-being: current state and trends: findings of the condition and trends working group. Island Press, Washington, DC

World Health Organisation (1990) Methylmercury. Environmental health criteria 101. World health organisation. Geneva, Switzerland

World Health Organisation (2002) Eutrophication and health. World health organisation regional office for Europe, and european commission, Luxembourg

Appendix References

Altman JC (1987) Hunter-gatherers today: an Aboriginal economy in north Australia. Australian Institute of Aboriginal Studies, Canberra

Altman J (2003) People on country, healthy landscapes and sustainable Indigenous economic futures: The arnhem land case. The drawing board: an Australian review of public affairs 4, 2. School of Economics and Political Science, University of Sydney. pp 65–82

Bollhöfer A (2012) Stable lead isotope ratios and metals in freshwater mussels from a uranium mining environment in Australia's wet-dry tropics. Appl Geochem 27:171–185

Bollhöfer A, Brazier J, Ryan B, Humphrey C, Esparon A (2011) A study of radium bioaccumulation in freshwater mussels, *Velesunio angasi*, in the Magela Creek catchment, Northern Territory, Australia. J Environ Radioact 102:964–974

Director of National Parks (2007) Kakadu national park management plan 2007–2014. Commonwealth of Australia, Canberra

Doering C, Bollhöfer A (2014) Radionuclide transfer to terrestrial bush foods. In: eriss research summary 2012–13. Supervising Scientist Report 205, Supervising Scientist, Darwin

International Commission on Radiological Protection (ICRP) (1995) Age-dependent doses to members of the public from intake of radionuclides: Part 5 Compilation of ingestion and inhalation dose coefficients. ICRP Publication 72, Annals of the ICRP 26(1)

International Commission on Radiological Protection (ICRP) (2007) The 2007 Recommendations of the International Commission on Radiological Protection. ICRP Publication 103, Annals of the ICRP 37(2–4)

Jeffree RA, Markich SJ, Brown PL (1995) Australian freshwater bivalves: their applications in metal pollution studies. Australas J Ecotoxicol 1:33–41

Johnston A (1987) Radiation exposure of members of the public resulting from operations of the Ranger Uranium Mine. Technical memorandum 20, Supervising Scientist for the Alligator Rivers Region, AGPS, Canberra

Markich SJ, Jeffree RA (1994) Absorbtion of divalent trace metals as an analogue of calcium by Australian freshwater vivalves: an explanation of how water hardness reduces metal toxicity. Aquat Toxicol 29:257–290

Martin P (2000) Radiological impact assessment of uranium mining and milling. PhD thesis. Queensland University of Technology, Brisbane

Martin P, Hancock GJ, Johnston A, Murray AS (1998) Natural-series radionuclides in traditional north Australian aboriginal foods. J Environ Radioact 40(1):37–58

Mitchell N, Pérez-Sénchez D, Thorne MC (2013) A review of the behaviour of U-238 series radionuclides in soils and plants. J Radiol Prot 33:R17–R48

Murray AS, Johnston A, Martin P, Hancock G, Marten R, Pfitzner J (1993) Transport of naturally occurring radionuclides by a seasonal tropical river, Northern Australia. J Hydrol 150:19–39

Pettersson HBL, Hancock G, Johnston A, Murray AS (1993) Uptake of uranium and thorium series radionuclides by the waterlily, Nymphaeaviolacea. J Environ Radioact 19:85–108

Pöschl M, Nollet LML (2007) Radionuclide concentrations in food and the environment. CRC Press, Taylor & Francis Group

Ramsar Convention (2012). Wetland tourism: Australia—Kakadu Ramsar Site. A Ramsar case study on tourism and wetlands. http://www.ramsar.org/tourism

Ranger Uranium Environmental Inquiry (RUEI) (1977) Ranger uranium environmental inquiry 2nd report. Canberra, Australian Government Publishing Service

Ryan B, Martin P, Iles M (2005) Uranium-series radionuclides in native fruits and vegetables of Northern Australia. J Radioanal Nucl Chem 264(2):407–412

Sauerland C, Martin P, Humphrey C (2005) Radium 226 in Magela Creek, Northern Australia: application of protection limits from radiation for humans and biota. Radioprotection 40(Suppl 1):451–456

Supervising Scientist (2012). Annual Report 2011–2012. Supervising scientist, Darwin. Australia

van Dam R, Humphrey C, Martin P (2002) Mining in the Alligator rivers region, Northern Australia: assessing potential and actual effects on ecosystem and human health. Toxicology 181–182:505–515

Healthy Wetlands, Healthy People: Mosquito Borne Disease

Scott Carver, David P. Slaney, Paul T. Leisnham and Philip Weinstein

Abstract We evaluate the links between wetland breeding mosquitoes (Diptera: Culicidae), vector-borne disease transmission, human incidence of disease and the underlying mechanisms regulating these relationships. Mosquitoes are a diverse taxonomic group that plays a number of important roles in healthy wetlands. Mosquitoes are also the most important insect vectors of pathogens to wildlife, livestock and humans, transmitting many important diseases such as malaria, West Nile virus, and Ross River virus. Mosquitoes interact with a variety of invertebrates and vertebrates in complex communities within wetlands. These interactions regulate populations of key vector species. Healthy wetlands are characterized by intact wetland communities with increased biodiversity and trophic structure that tend to minimize dominance and production of vector mosquito species, reservoir host species and minimize risk of disease to surrounding human and animal populations. In a public health paradigm, these natural ecological interactions can be considered a direct ecosystem service—natural mitigation of vector-borne disease risk. Anthropogenic disruptions, including land-use, habitat alterations, biodiversity loss and climatic changes can compromise natural ecological processes that regulate

S. Carver (✉)
School of Biological Sciences, University of Tasmania,
Private Bag 55, Hobart, TAS 7001, Australia
e-mail: scott.carver@utas.edu.au

D. P. Slaney
Institute of Environmental Science and Research Ltd,
Wellington, New Zealand and Barbara Hardy Institute,
University of South Australia, Adelaide, SA 5001, Australia
e-mail: david.slaney@unisa.edu.au

P. T. Leisnham
Ecosystem Health and Natural Resource Management,
Department of Environmental Science and Technology, University of Maryland,
1459 Animal Sciences Bldg, College Park, MD 20742, USA
e-mail: leisnham@umd.edu

P. Weinstein
School of Biological Sciences, The University of Adelaide, Adelaide 5005, Australia
e-mail: philip.weinstein@adelaide.edu.au

© Springer Science+Business Media Dordrecht 2015
C. M. Finlayson et al. (eds.), *Wetlands and Human Health,* Wetlands: Ecology,
Conservation and Management 5, DOI 10.1007/978-94-017-9609-5_6

mosquito populations and have severe human health and economic implications. Maintenance of healthy wetlands is likely to be beneficial for human and ecosystem health, and more cost effective and sustainable than chemical control of vector species.

Keywords Competition · Predation · Ecosystem · Biodiversity · Aquatic interactions · Community · Climate · Salinity · Mosquito-borne disease · Vector · Ross River virus · Malaria · West Nile virus · Filariasis

Introduction

Globally, wetlands are subject to a variety of natural and anthropogenic threats, of which there are multiple consequences to ecosystems within and surrounding these wetlands. In this chapter we focus on the ecological effects of wetland disruption and address their health related impacts relating to mosquito-borne disease. Mosquitoes (Diptera: Culicidae) are an important biological component of wetlands (Dale and Knight 2008). As a taxonomic group, mosquitoes exhibit diverse life histories and many mosquito species utilize wetlands as oviposition and larval development sites (Dale and Knight 2008; Gilbert et al. 2008). Within wetlands, mosquito larvae occupy a variety of species specific ecological niches as predators, grazers or filter feeders. The larvae of many mosquito species are relatively inferior competitors with other aquatic fauna and are susceptible to predation (Juliano 2007) and, as such, form an important component of wetland food webs. As a general rule, disruptions to wetlands that consequently result in enhanced mosquito-borne disease tend to be those that negatively impact competitive or predatory regulation of mosquito populations and, thus, erode ecosystem services that would otherwise benefit human health (Horwitz and Finlayson 2011).

In this chapter we emphasize the importance of ecological processes that regulate mosquito populations and communities in wetlands, and how changes to mosquito ecology can translate to disease transmission. We broadly review the literature and present specific case studies that relate pathogen/disease information to wetland and mosquito information. We focus on wetlands that naturally contain freshwater, under the following definition: a permanent, semi-permanent or ephemeral body of water, fed primarily by rainfall or groundwater hydrology, which is either natural or anthropogenic in origin. We describe the main anthropogenic sources of mosquito regulation and refer the reader to other sources of information (e.g., Dale and Knight 2008), but do not make this a focus of the chapter. We also exclude discussion on coastal tidal salt marshes, which can vary considerable in their ecology to freshwater wetlands (Dale and Knight 2008, 2012). This chapter shows that by understanding the ecological determinants of mosquito populations in wetlands, and consequently human disease incidence, wetland management that promotes natural mechanisms of control are likely to be more sustainable and economically beneficial than management involving chemical control.

Wetlands and Mosquitoes

Lifecycle

Most shallow aquatic habitats, including natural and constructed wetlands (e.g., Fig. 1), provide suitable habitat for a variety of mosquito species. Adult female mosquitoes oviposit on the surface of existing water bodies or on substrate (e.g., emergent vegetation, ground depressions) that will likely be inundated with water (Fig. 2). Mosquito larvae hatch and develop through four larval instars while feeding on microorganisms (bacteria, fungi, and algae) and detritus in the water column and on plant surfaces (Merritt et al. 1992) (Fig. 2). After a brief non-feeding pupal stage, winged adults emerge (Fig. 2). Adults usually mate soon after hatching and feed on nectar, but females of all but a few species also require a blood-meal to provide proteins and nutrients for egg development. Females often feed on multiple hosts because they either get insufficient blood required for oviposition from one host or survive long enough to lay multiple batches of eggs. The blood feeding habit of female adults renders them prone to acquiring pathogens and transmitting diseases when they feed on an infected host and then bite a non-immune human. Although only adult females directly interact with humans, the distribution and abundance of mosquito species are mainly regulated by ecological processes at the egg (e.g., desiccation) and larval (e.g., competition and predation) stages (Juliano 2009). Environmental factors that indicate favorable oviposition habitats to adult females also indirectly regulate local mosquito distributions and abundances. Because wetlands represent extensive potential mosquito development sites and population dynamics are driven by processes acting at the immature stages, studies of mosquito communities in both natural and constructed wetlands have been a focus of considerable

Fig 1 A constructed wetland in a residential development in Centreville, Maryland, USA. This wetland has been designed to permanently retain water. It is an example of one of the most accepted and effective best management practices for treating stormwater runoff and removing harmful nutrients. (© Paul Leisnham)

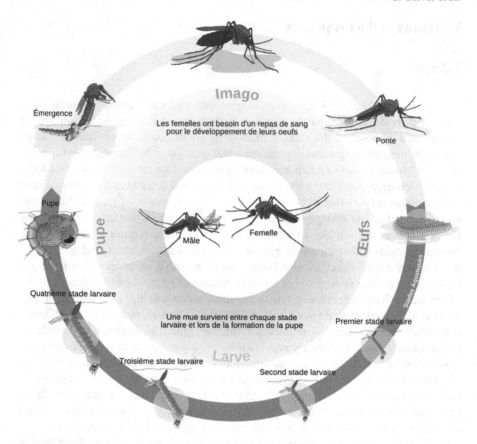

Fig 2 Generalized lifecycle for a *Culex* sp. mosquito representing Adult (Imago), egg, larval and pupal stages. Mariana Ruiz Villarreal, Wikimedia Commons: http://commons.wikimedia.org/wiki/File:Culex_mosquito_life_cycle_en.svg

research in pest control, water management, and public health (e.g., Knight et al. 2003; Rey et al. 2012).

As with most wetland insects, the life histories of mosquitoes are strongly influenced by the flooding regimes and the permanence of habitats (e.g., Washburn 1995; Batzer and Wissinger 1996; Wellborn et al. 1996; Juliano 2009). Insect life histories have commonly been separated into two broad categories based on their colonization of wetlands during and after flooding (Batzer and Wissinger 1996). This categorization is also relevant to mosquitoes. The first category of wetland colonization involves the oviposition of desiccation-resistant eggs on substrates that are later inundated with water. This strategy is characteristic of univoltine *Aedes* species in temperate regions that overwinter at the egg stage, such as those species that oviposit in ground depressions that later become ephemeral snow-pools (e.g., *Aedes communis* and *Aedes punctor* of the northern United States and Canada), or species that oviposit on aquatic vegetation above the water line in more permanent pools (e.g., *Aedes abserratus*). Although these species are thought to be univoltine and are characterized by peaks in abundance early in the season, some frequently reappear more than once during a single breeding season. This reappearance may represent a percentage of the overwintering population that requires more than one flooding for complete egg hatch or a staggering of egg maturation within clutches so that some eggs remain viable over several wet-dry cycles. Alternatively, late season individuals may also represent a second generation. In addition to univoltine species, there are numerous multivoltine species that utilize multiple flooding events of ephemeral wetlands each season by laying desiccation resistant eggs in ground-pool depressions. These species are commonly referred to as floodwater mosquitoes (e.g., *Aedes vexans*, *Aedes alboannulatus* and *Aedes camptorhynchus*) and undergo accelerated larval development and pupation within 4–5 days of egg hatch if water temperatures are favorable.

The second category of colonization involves adult immigration and oviposition on water surfaces. These mosquitoes have eggs that need to be continually moist and hatch with a few days after oviposition (e.g., *Culex australicus* and *Anopheles annulipes*). In temperate regions, many *Culex* and *Anopheles* mosquitoes overwinter at the adult sage and then colonize wetlands at the start of the season. Breeding is continuous; thus, all instars are represented in a wetland at any one time. Often these species utilize permanent wetlands that have considerable organic matter and thus many species are tolerant to high organic content (e.g., *Culex pipiens*, *Culex quinquefasiatus* and *Culex australicus*).

Because immature mosquitoes are restricted to the specific wetland of maternal oviposition, there is strong selection on adult females to identify favorable habitats (Laird 1988; Blaustein and Chase 2007). Several studies have demonstrated that physiochemical conditions of wetlands cue either positive or negative oviposition responses in adult female mosquitoes or that species vary considerably in their responses (Clements 1999). These include humidity gradients (Kennedy 1942), vegetation cover (Orr and Resh 1992), and food (Mian and Mulla 1986; Blaustein and Kotler 1993). Additionally, several species of mosquito have been observed to avoid ovipositing in wetlands with predators (Chesson 1984; Petranka and Fakhoury 1991; Angelon and Petranka 2002; Spencer et al. 2002; Kiflawi et al. 2003; Silberbush and Blaustein 2008; Walton et al. 2009) or competitors (Orr and Resh 1992; Blaustein and Kotler 1993; Edgerly et al. 1998; Kiflawi et al. 2003; Mokany and Shine 2003). In addition to conspecifics, common competitors of mosquitoes include a variety of other co-occurring mosquito species and other invertebrates (e.g., non-mosquito insects (Chironomidae, several Corixidae), molluscs (Pulmonate snails), anuran larvae (Ranidae, Hylidae), amphipods, and cladoceran and copepod zooplankton). Considerable research has attempted to identify specific chemical compounds and microbial biota related to oviposition to use both as attractions in oviposition traps and deterrents (Clements 1999; Silver 2008).

Wetlands and Mosquito-Borne Disease

There are a number of vector-borne diseases around the world that are transmitted by wetland breeding mosquitoes (Dale and Knight 2008). Some of the mosquito species involved in disease transmission also utilize other breeding habitats such as containers, but here we focus upon non-tidal wetland environments. Specifically, we focus on Ross River virus (RRV, Togoviridae: *Alphavirus*), malaria and West Nile virus (WNV, Flaviviridae: *Flavivirus*), which are featured in case studies throughout this chapter. We also describe filariasis, which is also mosquito-borne and associated with wetland environments, but not featured in our case studies.

Ross River Virus Ross River virus (RRV, Togoviridae: *Alphavirus*) is a mosquito-borne pathogen, endemic to Australia and Papua New Guinea (Fig. 3), and of significant public health concern. Epidemiologically RRV is Australia's most significant (causing the most human illness) mosquito-borne pathogen, resulting in approximately 5000 human notifications of polyarthritis, the clinical manifestation of RRV infection, annually (Russell 2002). The most common presenting symptoms of people infected with RRV are arthralgia, low-grade fever, fatigue, and a maculopapular rash. Although 55–75 % of cases of RRV infection are asymptomatic (Aaskov et al. 1981; Russell et al. 1998; Harley et al. 2001), the disease can be responsible for considerable morbidity, particularly with debilitating joint pains that can persist for months.

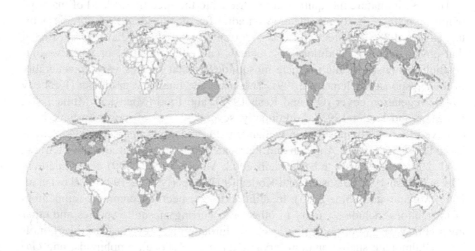

Fig 3 Map showing countries where cases of Ross River virus (*upper left*), Malaria (*upper right*), West Nile virus (*lower left*), and Lymphatic filariasis (*lower right*) occur. Data sources: RRV—Lau et al. (2012) and Harley et al. (2001); Malaria—WHO (http://gamapserver.who.int/mapLibrary/) and CDC (http://wwwnc.cdc.gov/travel/yellowbook/2014/chapter-3-infectious-diseases-related-to-travel/travel-vaccines-and-malaria-information-by-country/azerbaijan#seldyfm533); Lymphatic filariasis—WHO (http://gamapserver.who.int/mapLibrary/); *WNV*—World Animal Health Information Database (http://www.oie.int/wahis_2/public/wahid.php/Diseaseinformation/Diseasedistribution-map) and Pubmed searches (http://www.ncbi.nlm.nih.gov/pubmed)

This arbovirus cycles enzootically between mosquito vectors (which primarily develop in freshwater wetlands and saltmarsh habitats) and mammalian reservoir hosts. Human cases often occur in or near areas with abundant vector and host populations, where epizootics likely lead to spillover and epidemics (e.g., Lindsay et al. 1996). There are three principal vectors of RRV across Australia (*Culex annulirostris*, *Aedes camptorhynchus* and *Aedes vigilax*), but other species can also be involved (Harley et al. 2001; Russell 2002). *Culex annulirostris* breeds in freshwater wetlands and contributes to transmission across the continent, *Aedes camptorhynchus* is halotolerant and is a dominant vector in southern Australia, and *Aedes vigilax* is also halotolerant replacing *Aedes camptorhynchus* north of its distribution (Harley et al. 2001; Russell 2002). Both of these species breed primarily in saltmarsh habitats, but also occur in inland riverine and wetland areas where there is a brackish influence. Across the continent macropod marsupials are generally considered the principal reservoir hosts, although other marsupials (and to a lesser extent, placental mammals) also have a role in transmission (Boyd et al. 2001; Boyd and Kay 2001; Harley et al. 2001; Russell 2002; Kay et al. 2007; Carver et al. 2009c). Activity of reservoirs around wetlands influences RRV outbreaks. For example, RRV outbreaks in Tasmania have been linked to activity of macropods around flooded pastures where *Aedes camptorhynchus* have become abundant (Russell 2002).

Malaria Malaria is one of the most prevalent vector-borne diseases in the world, causing an estimated 225 million cases and 781,000 deaths in 2009 (WHO 2010), and claiming the lives of more children than any other infectious disease. Symptoms of malaria include an acute febrile illness, with the most common presenting symptoms being fever, headache, chills and flu-like illness. The disease is endemic in a number of countries across the tropics and sub-tropics, in particular Africa (Fig. 3). There are four malaria parasites (eukaryotic protists of the genus *Plasmodium*) that infect humans: *Plasmodium falciparum*, *Plasmodium vivax*, *Plasmodium ovale*, and *Plasmodium malariae* (the first two being the most common). A fifth malaria parasite (*Plasmodium knowlesi*, monkey malaria) has recently emerged as a public health threat with human cases being reported from Southeast Asia (Sabbatani et al. 2010). The primary species responsible for severe symptoms and human mortality is *Plasmodium falciparum*, and we focus on this species in all discussion that follows.

Malaria (*Plasmodium falciparum*) cycles between humans and primarily *Anopheles* mosquitoes. It is host, but not vector, species specific. However, *Anopheles gambiae* sensu stricto is considered one of the most efficient vectors of malaria (Sinka et al. 2010). This species is highly anthropophilic in its host feeding preference and commonly selects human structures as rest sites during the day (Sinka et al. 2010). Incidence of malaria has historically been associated with the presence of wetlands (swamps and marshes). Consequently management to reduce human incidence has included the deliberate draining of these habitats. Both historical and contemporary incidence of malaria is more often associated with water development projects, agriculture, urbanisation, deforestation, and other human modifications to the environment that create anthropogenic wetlands (Gratz 1999; Patz et al. 2004; O'Sullivan et al. 2008).

West Nile Virus West Nile virus (WNV, Flaviviridae: *Flavivirus*) is a mosquito-borne virus, with a distribution spanning five continents (Fig. 3) and is of significant health concern to humans. Infection with WNV can be asymptomatic, result in West Nile fever (a mild febrile illness), or in severe cases West Nile encephalitis (a neurodegenerative disease), which can be fatal (Smithburn and Jacobs 1942; Tsai et al. 1998). However, the vast majority (~80–90%) of WNV cases are asymptomatic (Steinman et al. 2003). Increased attention has been given to the ecology and epidemiology of WNV, since its introduction to North America in 1999 and rapid spread from the east to west coast (Petersen and Hayes 2004). Sequence data indicates WNV in North America was most closely related to a strain found in Israel in 1998. In the U.S. over 35,000 people have been infected with WNV, leading to over 1500 deaths (CDC 2013).

This virus is primarily transmitted by *Culex* spp. mosquitoes, amplified by competent avian hosts and often associated with wetlands (Hubalek and Halouzka 1999; Kilpatrick et al. 2006b; Ezenwa et al. 2007; Kilpatrick et al. 2007; Reiter 2010). WNV can also be transmitted to a variety of mammal species, although they are less competent reservoirs than birds (Kilpatrick et al. 2005; Hamer et al. 2008). An important vector for WNV within North America and Europe is *Culex pipiens* (Fonseca et al. 2004b), although vector species vary geographically (Kilpatrick et al. 2007). WNV exhibits extreme temporal and spatial variability (Kilpatrick et al. 2010), and a number of studies have demonstrated this is related to host community structure (Ezenwa et al. 2006; Kilpatrick et al. 2006b; Swaddle and Calos 2008; Allan et al. 2009). Vector feeding behavior, and shifts in feeding behavior, also appear important determinants of the spatial and temporal distribution of human cases (Fonseca et al. 2004b; Fonseca et al. 2004a; Spielman et al. 2004; Kilpatrick et al. 2006a).

Filariasis A fourth example of mosquito-borne disease associated with wetlands is filariasis. This is a chronic parasitic disease caused by infection from filarial nematode worms. There are three types of human filarial disease, defined by the region of the body that the nematodes invade. Only lymphatic filariasis, which occupies the lymphatic system, is vectored by mosquitoes. The disease is endemic in a number of countries across the tropics and sub-tropics of Africa, Asia, Western Pacific, and Central/South America (Fig. 3), with approximately 120 million people infected (CDC 2011; Taylor et al. 2010). Of those, about two thirds are asymptomatic, while the remainder show clinical signs of the disease, such as hydrocoele, lymphoedema, and elephantiasis (CDC 2011; Taylor et al. 2010).

The lymphatic filariasis causing nematodes, *Wuchereria bancrofti*, *Brugia malayi*, and *Brugia timori*, are transmitted by a wide range of mosquito species within the *Aedes*, *Anopheles*, *Coquillettidia*, *Culex* and *Mansonia* genera (CDC 2011). The specific vectors involved in transmission vary by geography and nematode species, with a number of the mosquito species proliferating in wetland areas where the disease is endemic. However, the disease is predominantly associated with wetlands that are anthropogenic in origin (including areas of poor surface drainage, irrigation schemes, large numbers of container habitats, and lack of sanitation) rather than the presence of natural wetlands (Erlanger et al. 2005). For example, in Timor Leste, 95% of lymphatic filariasis cases are caused by *B. timori*, vectored by *Anopheles barbirostris*. This mosquito species breeds in water from fresh water springs but has a particular affinity to irrigated rice-paddy fields (Sudomo et al. 2010). Adult nematodes in the lymphatic system produce microfilariae that migrate into the lymph and blood channels. These microfilariae often concentrate in the blood vessels with a circadian rhythm that matches the bitting patterns of the local mosquito vectors (Taylor et al. 2010). It is within the mosquito that the microfilariae continue their life cycle before being transmitted to humans as larvae when the mosquito bites again.

Regulation of Mosquitoes in Wetlands

There are a variety of ecological processes by which mosquitoes are regulated in wetlands (Fig. 4). In a public health sense, ecological processes that regulate mosquito vectors of infectious diseases to humans can be considered ecosystem services (Patz and Confalonieri 2005; Horwitz and Finlayson 2011). Here, we briefly consider the main ecological processes that regulate mosquitoes in wetlands, before delving into more detailed assessments of the effects of wetland disruption and health of wetlands on vector mosquito abundance and vector-borne diseases.

Factors that regulate wetland permanence (ephemerality), structure (vegetation and physical characteristics) and nutrient characteristics are perhaps the most conspicuous determinants of mosquito abundance, human-mosquito contact and vector-borne disease incidence (Wellborn et al. 1996, e.g., Pinder et al. 2005; Sanford et al.

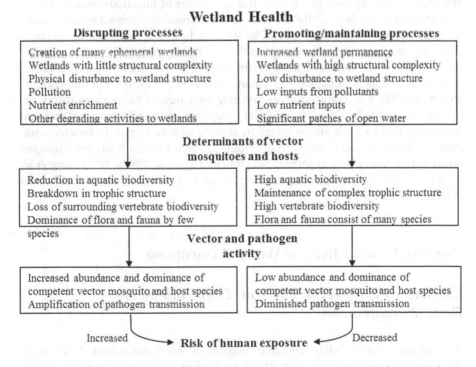

Fig 4 General linkages between wetland health and processes governing abundance and dominance of vector mosquitoes, abundance and dominance of reservoir hosts, pathogen activity and human exposure risk

2005; Juliano 2009; van Schie et al. 2009; Carver et al. 2010). In general, evidence suggests the abundance of vector mosquitoes is lower in wetlands with more open and deeper water (which is likely to exhibit more surface motion/disturbance), with relatively sparse or patchy macrophytes and emergent vegetation, and not rich in nutrients (such as nitrogen) (Russell 1999; Thullen et al. 2002; Arav and Blaustein 2006; Gingrich et al. 2006; Munga et al. 2006; Carver et al. 2009b; Juliano 2009; Mian et al. 2009). These factors generally promote complex aquatic communities, where predators and competitors of mosquito larvae are abundant. Greater motion of the water surface in more open wetlands tends to negatively influence mosquito abundance because larvae of many vector species must swim to the water surface to breathe, and hence require relatively still water.

However, because many vectors of mosquito-borne disease are capable of rapidly colonizing bodies of water, the appearance of ephemeral wetlands is also an important source of mosquito abundance (Gingrich et al. 2006; Carver et al. 2009b). As a general phenomenon, ephemeral wetlands serve as significant sources of vector mosquito production compared to permanent wetlands because they are deficient in predatory and competition based regulatory processes (Spencer et al. 1999; Chase and Knight 2003; Schafer et al. 2006; Juliano 2009; Carver et al. 2010). Hence, there is often a capacity of mosquitoes to achieve greater densities and per capita rate increases within ephemeral wetlands (Chase and Knight 2003).

The factors we have briefly described above are important in regulating the abundance of mosquitoes in wetlands that are vectors of infectious disease. This is important because the potential for pathogen transmission is related to vector mosquito abundance (e.g., Carver et al. 2009a). It is, however, important to acknowledge that additional phenomena exist in the area surrounding wetlands that regulate the prevalence of infections among vector species, and consequently exposure risk of humans to infectious pathogens. In particular, these are factors that influence the structure of the host community surrounding wetlands. In many cases pathogens that are transmitted by mosquitoes (with exceptions, such as Dengue 1–4 and malaria) are either zoonotic and/or spread by catholic vectors. Hence the relative abundance of hosts that act as amplifying or dilution agents (collectively encompassed under the dilution effect hypothesis; Ostfeld and Keesing 2000a, b; Keesing et al. 2006), as well as vector feeding preference, are important components of human disease exposure risk.

Negative Health Effects of Wetland Disruption

Biodiversity Loss, Vector and Host Communities, and Disease Proliferation

As outlined in the preceding section, ecological processes that negatively influence abundance of mosquito larvae in wetlands tend to be those that (1) promote community complexity and consequently increased interactions between mosquitoes and aquatic competitors and predators, or (2) increase physical disturbance of the water surface (Fig. 4). Accordingly, disruptions to wetlands that influence these features tend to be those that are likely to promote abundant populations of mosquito larvae and the capacity of the vector population to transmit pathogens (Fig. 4 and 5). Additionally, disruptions to wetlands and their host communities are also important in pathogen transmission. Disruption of the host community can result in increased

Fig 5 Ephemeral wetlands in the wheatbelt of southwestern Australia: **a** healthy wetland composed of rich aquatic community and few vector (*Aedes camptorhynchus*) mosquitoes; **b** wetland heavily impacted by rising groundwater tables and salinity with poor aquatic community richness and many *Ae. camptorhynchus*. Pictures courtesy of Scott Carver

pathogen transmission and disease incidence via mechanisms that include: local exclusion or reduction in abundance of dilution hosts, increased abundance of amplification hosts, or changes in host community structure that results in shifts in vector feeding behaviour toward amplification hosts or humans. In the broadest sense, these ecosystem disruptions can erode ecosystem services that are otherwise beneficial in reducing human mosquito-borne disease exposure risk. We present two case studies that illustrate concepts of how wetland disruption can influence vector mosquito abundance and transmission of vector-borne pathogens. We illustrate how (1) salinity induced disruption of wetlands influences vectors, hosts, determinants of vector abundance and the potential for pathogen transmission, and (2) how human changes in land-use (athropogenic availability of wetlands) and climate influence disease transmission.

Case Study: Dryland Salinity and the Ecology of RRV in Western Australia

In the southwest of Western Australia (WA) RRV is primarily transmitted by the salt tolerant mosquito *Aedes camptorhynchus*. This vector is common to coastal saltmarsh habitats, but rarely occurs in freshwater wetlands, even though the larvae develop well in fresh water (Barton and Aberton 2005; van Schie et al. 2009). Available evidence suggests that *Aedes camptorhynchus* larvae are restricted to brackish and saline environments because they are inferior competitors and susceptible to predation in freshwater environments (Carver et al. 2010). Accordingly, incidence of RRV in southwest WA predominates around coastal saltmarsh zones, where *Aedes camptorhynchus*, mammalian reservoir hosts and humans are abundant (Lindsay et al. 1996; Lindsay et al. 2005).

The inland southwest of WA (the wheatbelt) is substantially affected by secondary salinisation, as a consequence of land clearing and rising groundwater tables bringing salt stored in the regolith to the soil surface (McKenzie et al. 2003). Currently more than 1 million hectares are affected and salinisation is expected to expand two to four fold by 2050 (George et al. 2006). This environmental change has resulted in salinity induced disruption of many wetlands in this region, with dramatic consequences for aquatic fauna and wetland structure (e.g., Bunn and Davies 1992; Halse et al. 2003; Cale et al. 2004; Pinder et al. 2005; Carver et al. 2009b, 2010) (Fig. 5).

Of particular relevance to the ecology of RRV, increasing salinization of the wheatbelt has resulted in greater regional abundance and distribution of *Aedes camptorhynchus* (Lindsay et al. 2007; Jardine et al. 2008a; Carver et al. 2009b). The likely underlying mechanism for this increased vector abundance and distribution has been a combination of greater area of standing water for *Aedes camptorhynchus* breeding and release from aquatic competition and predation in water bodies contaminated with salt (Carver et al. 2009d; Carver et al. 2010) (Fig. 5).

This increased vectorial potential for RRV transmission is broadly consistent with ecosystem distress syndrome (Jardine et al. 2007), as originally described by Rapport et al. (1985).

Beyond the direct disruptive effects of salinity on wetlands and their species composition in the region, there is potential for salinity to have other influences on the ecology of RRV. In particular, mammalian hosts (particularly marsupials) around wetlands are important components of zoonotic RRV transmission. Their distribution is influenced by habitat characteristics in the environment, which are disrupted by salinity (McKenzie et al. 2003, 2004). Surveys of mammal community structure were undertaken in the region and, surprisingly, mammal community structure was relatively unaffected by environmental salinity (Carver et al. 2009a; Carver 2010). Subsequent modeling of potential RRV transmission dynamics revealed that the major determinant of RRV transmission in this region was abundance of *Aedes camptorhynchus* (Carver et al. 2009a).

Ecological conditions in the southwest of WA suggest that one might expect to see more human cases of RRV in areas disrupted by salinization (Carver et al. 2009a). However, to date no such relationship has been detected (Jardine et al. 2008b). Currently, there are relatively few cases of RRV in this region, and it appears human cases are more closely related to coastal epidemic activity, and subsequent dispersal of infected humans, *Aedes camptorhynchus*, or hosts inland (Jardine et al. 2008b; Carver et al. 2009a). It is possible that an enzootic cycle for RRV could establish in salinized areas, leading to increased human incidence. However, it is currently unclear if host community structure and seasonal environmental conditions within areas of the region would be permissive for establishment of such a cycle. More probable is that the disruption of wetlands by salinity has had, and continues to have, an amplifying effect on vector mosquito abundance and that the potential for higher RRV incidence rates into the future.

Case Study: Climate, Wetlands and Health

Links between climate change and mosquito-borne disease have received considerable attention over the past 10 years (Rohr et al. 2011). However, much of this attention to date has been speculative with little empirical support. It is widely acknowledged that climate change and variation may have an effect on the biology of vectors, biotic determinants of their abundance and hence alter seasonality, abundance and distributions (Githeko et al. 2000; Heft and Walton 2008; Reisen et al. 2008; Lafferty 2009; Russell 2009; Mordecai et al. 2013). Likewise climate change might affect hosts and pathogens, and a number of socio-economic factors can also affect vectors, hosts and pathogens (Hales et al. 1999; Patz et al. 2005; Dale and Knight 2008; DeGroote et al. 2008; Reiter 2008; Dobson 2009; Lafferty 2009; Russell 2009; Mills et al. 2010; Rohr et al. 2011). For example, climate does influence human behaviour and hence exposure risk, such as warmer weather leading to greater time spent outdoors (Tucker and Gilliland 2007). With this complex inter-relationship it can be difficult to directly measure and model the relationships between climate change and vector-borne disease exposure and effect.

A number of mosquito vector species utilize wetlands for breeding (see Regulation of mosquitoes in wetlands; Dale 1993; Dale and Knight 2008) and one would expect climatic changes that affect wetland size and availability in time and space will influence mosquito seasonality, abundance and distributions, both directly and via biotic determinants of abundance. Furthermore, a range of hosts involved in vector-borne disease dynamics will also be affected by wetland availability e.g. wetland birds, macropods, cattle, and this in turn may influence disease incidence. Climatic changes that bring about the expansion or reduction of wetland habitats, or changes in permanence, may thus alter future mosquito-borne disease incidence, such as WNV or RRV. However, it cannot be concluded that the mere expansion of wetland habitats will lead to increased disease incidence as complex biological inter-relationships come in to play. For example, in the western United States, WNV outbreaks occur following drought years, where predators of wetland-breeding mosquitoes have been previously eliminated by the dry conditions and allow mosquito proliferation in the subsequent wetter year (Chase and Knight 2003; Landesman et al. 2007; DeGroote et al. 2008; Lafferty 2009).

Russell (2009) suggested that for Australia, climatic changes that extend the permanence of seasonal wetlands in northern Australia will likely enhance cycles that sustain Murray Valley encephalitis virus and Kunjin virus across the northern top of Western Australia to Queensland. Counter to this, the projected drier spring conditions in south-eastern Australia may reduce virus activity by affecting bird and mosquito breeding through loss of wetland habitat (Russell 2009). Again, due to the complex biological inter-relationships these predictions are not clear cut. Russell (2009) points out that heavy rainfall events can be detrimental for *Culex* larvae via habitat flushing, while for arid grasslands, the same event may result in extensive activity of floodwater *Aedes* species and subsequently *Culex annulirostris* that utilises newly vegetated and persisting habitat. Greater availability of habitat for *Culex annulirostris* may also lead to an increase in RRV activity. However, based on current climate models it is likely there will be a decrease in rainfall over most of Australia including the far north (CSIRO and the Bureau of Meteorology 2011).

As described above, there are significant challenges in drawing conclusive links between climate change and vector-borne disease, associated with the complexities of these systems and data requirements in evaluating links. Lafferty (2009) correctly surmised that, although the globe has warmed significantly, there is little current evidence that this has increased incidence of infectious diseases. More accurate projections are that there will be changes in the geographic range of infectious diseases with little net increase in area (Mordecai et al. 2013). Lafferty (2009) also concluded that non-climatic factors (such as effective use of mosquito nests, socio-economic factors, human behavior, changing land-use, etc.) may overshadow the effects of climate on infectious diseases. Furthermore, a variety of sources cite land-use change, the primary force altering the environment in the twenty first century, as likely to be a greater driver for the emergence and re-emergence of many infectious diseases, including mosquito-borne diseases (e.g., Patz et al. 2004; Jones et al. 2008; Olson et al. 2010; Slaney et al. 2010; see our other case studies). Thus, the relative impact of climate change on mosquito-borne diseases must be evaluated within the context in which a particular mosquito-borne disease occurs.

Despite these significant challenges some links with climate change have been documented for several infectious diseases, mostly of which are vector-borne (Ostfeld 2009; Rohr et al. 2011 and references therein). Of these examples there is one that is both mosquito-borne, associated with wetlands and exhibiting increasing global incidence, Malaria (Pascual et al. 2006; Pascual et al. 2008; Alonso et al. 2011). Authors have principally been able to demonstrate links between climate change and malaria because of high quality long-term climate, vector and disease data, thorough understanding or the biology of the system, and use of models that incorporate both extrinsic (climate) and intrinsic (biology of parasite, hosts and vectors) complexities of the system (Ostfeld 2009). Such detailed knowledge is not available for most other mosquito-borne pathogens associated with wetlands.

In terms of human mortality, malaria causes the greatest disease burden of any mosquito-borne disease. *Anopheles gambiae* sensu stricto (the most efficient malaria vector) is associated with semi-permanent, often man-made, or frequently disturbed or ephemeral habitats in agricultural areas, such as rice fields and irrigated areas (Sinka et al. 2010; Yadouleton et al. 2010). These habitats are frequently deficient in aquatic predators and competitors of mosquitoes. *Anopheles gambiae* also has a short larval development period, is long lived and highly anthropophilic. The combination of these factors likely enables many larvae to develop to adulthood before competitors and predators colonize water bodies, to respond rapidly to local increases in daily temperature, provides sufficient time for the parasite to develop within the vector, and be transmitted to humans.

Pascual et al. (2006) concluded that in east African highlands, where malaria incidence has increased since the 1970s, that temperature has increased and vector mosquito population dynamics increase in an exponential fashion with temperature. Consequently, the size of malaria epidemics in these regions has increased since 1970 and cases exhibit a highly non-linear response to this changing climate (Pascual et al. 2008; Alonso et al. 2011). Comprehension of the mechanistic links between climatic changes and exacerbated malaria incidence was advanced further by Paaijmans et al. (2009, 2010), who evaluated the interacting impacts of changes in average temperature and daily temperature fluctuations. In areas where malaria transmission is expected to expand (highlands, with cooler mean temperatures), fluctuations lead to increased parasite development rates and hence malaria transmission. Conversely parasite development rates and malaria transmission had the opposite pattern in lowlands, which have warmer mean temperatures and where malaria transmission is expected to contract. Thus, anthropogenic creation of wetlands and modification of wetlands by anthropogenic activities, particularly agriculture, has led to enhanced malaria incidence in Africa. When combined with anthropogenic induced climatic changes this has resulted in expanding incidence in African highlands and reduced incidence in lowlands.

Given complex inter-relationships and requirements for high quality data it is not surprising that there are few empirical studies that draw robust links between wetlands, biological processes within wetlands, mosquito-borne disease and climate change. As Dale and Knight (2008) point out, there are numerous papers that discuss the possible effects of climate change on mosquitoes and vector-borne diseases, but a lack of research linking the mechanistic components, such as trophic processes in wetlands. This highlights a fruitful area of research where gaps in our knowledge of these links should be investigated. In doing so we must bear in mind that no one study is likely to be of generic use and any predictive models will need to take into account regional variation in all model parameters as well as the context in which particular mosquito-borne diseases occur.

Positive Effects of Healthy Wetlands

Healthy Wetlands and Ecological Control of Vectors

Evaluation of how wetland disruption can promote incidence of mosquito-borne disease is important to understand in context of elucidating ecological mechanisms that underlie pathogen transmission. It is also important to evaluate healthy wetlands so as to better comprehend how to mitigate disease incidence and promote human and environmental health concurrently. In this section we focus on examples of healthy wetlands and positive outcomes for reducing vector abundance and mosquito-borne disease risk; a 'double dividend' whereby enhanced or maintained ecosystem services of wetlands benefit human health outcomes (see Fig. 3 of Horwitz and Finlayson 2011 and "Wetlands as settings for human health—the benefits and the paradox"). We present two case studies that illustrate (1) appropriate construction and maintenance of wetlands and (2) how maintaining diverse host community composition around healthy wetlands is associated with reductions in pathogen prevalence among vector mosquitoes.

Case Study: Macrophytes, Aquatic Fauna, Wetlands and RRV in Brisbane, Queensland

The Brisbane metropolitan region of Queensland, Australia, represents an interesting example of wetland health, vectors and mosquito-borne disease. Brisbane is a large populous (>2 million people) city, spread over a floodplain and surrounding hills, with a humid subtropical climate. This combination of demographic and environmental factors means that there are numerous wetlands throughout the metropolitan region and a large local human population which can become infected by mosquito-borne diseases, such as RRV. Queensland has the highest rates of RRV incidence in Australia, likely owing to the combination of moist-warm climatic conditions and abundant populations of mosquitoes and marsupials. As such, cases of RRV around Brisbane generally follow rainfall patterns (Muhar et al. 2000). RRV is primarily transmitted within urban areas of Brisbane by *Culex annulirostris* (originating from wetlands) and around coastal areas by *Aedes vigilax* (originating from coastal saltmarsh areas, not the focus of this chapter; Muhar et al. 2000). A variety of studies have been undertaken around aspects of wetland health, abundance of *Culex annulirostris* and RRV incidence, although none specifically focus on the links across this broad topic.

Wetlands within suburban areas including Brisbane are valued for aesthetic, biodiversity, flood and water management, and recreational reasons, as well as for providing habitat for birds (Boully 1998; Greenway 2005; Verhoeven et al. 2006). Paradoxically, wetlands are also frequently viewed negatively for the associated nuisance of biting mosquitoes and exposure risk to vector-borne diseases (see Sect. 3; Dale 1993; Boully 1998; Dale and Knight 2008). With respect to RRV, wetlands have been identified as a risk factor associated with the disease, and outbreaks and clusters of RRV have occurred in residential areas where cases have been located close to saline and freshwater wetlands (Whelan et al. 1997; Russell 1999, 2002; Kelly-Hope et al. 2004). The spatial distribution of RRV cases is not uniform across the Brisbane metropolitan region, with high human infection rates noted in certain areas including those containing freshwater wetlands (Muhar et al. 2000). In support of public perceptions that freshwater wetlands lead to higher RRV incidence, Muhar et al. (2000) found positive relationships between wetland and bushland cover among Brisbane suburbs and cases of RRV. This analysis did include saltmarsh habitat, in addition to freshwater wetlands, but Muhar et al. (2000) did note that RRV cases associated with freshwater suburban areas were characterised by high wetland cover, vegetation density and abundance of *Culex annulirostris*.

While the study by Muhar et al. (2000) would suggest wetlands are bad for human health in suburban Brisbane, it is important to note that there was little information in this study on wetland health. This brings us to a key but obvious point that not all wetlands are equal. Different species of mosquitoes are adapted to proliferate in specific types of wetlands and the ecological condition or health of the wetland can influence mosquito abundance and thus human exposure risk. Dale (1993) lists a range of breeding habitats that different Australian mosquito pest species occur in, including wetlands, and points out that *Culex annulirostris* is also found in wetlands polluted by human-related activities. Given many mosquito species can proliferate in polluted nutrient rich habitats, maintaining the health of wetlands may help in limiting mosquito abundance. Furthermore, polluted or disrupted wetland ecosystems may also lead to the reduction or elimination of mosquito larval predators and/or changes in the macrophyte coverage, in turn providing ideal habitats for mosquito breeding (Greenway et al. 2003). The advantages of a balanced wetland ecosystem also holds true for constructed wetlands, if the wetlands is designed and managed to maximise macroinvertebrate predators and control macrophyte coverage (Russell 1999; Hurst et al. 2004; Greenway 2005).

Greenway et al. (2003) provided a useful study demonstrating how the health of wetlands, even constructed wetlands, play an important role in mosquito abundance in Queensland (the data was from areas beyond suburban Brisbane). Among constructed wetlands, the abundance of mosquito larvae was reduced in wetlands with high biodiversity of macrophytes and macroinvertebrates (Greenway et al. 2003). Conversely, mosquitoes were most abundant where constructed wetlands consisted of dense mats of grasses (Greenway et al. 2003). It was concluded that abundance of mosquito larvae can be reduced by planting a variety of macrophyte types, maintaining at least 30% open water and excluding monoculture by any dominant macrophyte species. Combined these factors are consistent with evidence that promoting healthy "rich" wetlands with appropriate physical structure is beneficial for abundance of aquatic predators and competitors of mosquito larvae, thus reducing the vectorial risk for disease transmission. Such a situation also promotes conditions suitable for wetland birds, which are generally incompetent hosts for RRV.

Currently there has been little research on constructing functioning wetlands and vector mosquito abundance. Dale and Knight (2008) in their review identified the need for research into the links between restoring wetland ecosystem function and mosquito proliferation, because inappropriate restoration of wetlands function may well bring about unintended effects of creating mosquito habitat. Managing the mosquito and RRV aspects of suburban Brisbane generally falls upon local and state government, and primarily involves larvicidal treatment of wetlands, spraying adulticides, and physical management of wetland structure (Dale 1993). Local and state governments are also involved in wetland conservation and management, although the Federal government has a role in the protection of wetlands of national environmental significance. In addressing these two aspects of wetland management there need not be a conflict as the promotion and protection of healthy wetlands may help limit the proliferation of mosquitoes and disease risk.

Case study: Wetlands, Avian Communities and the Ecology of WNV

WNV is an interesting system for evaluating links between wetland health and mosquito-borne disease because vectors for this pathogen are commonly associated with wetlands. Here we focus on studies of WNV transmission in North America. Transmission of WNV exhibits considerable spatial and temporal variability in North America, and there are a variety of vector mosquito species (i.e., Petersen and Hayes 2004). The most important vector species for WNV transmission in northeast and northcentral North America appear to be *Culex pipiens* and *Culex restuans* and, on occasion, *Culex salinarius* (Kilpatrick et al. 2005). Across southern North America *Culex quinquefasciatus*, *Culex nigripalpus* and *Culex erraticus* appear to be the most important (Turell et al. 2005; Cupp et al. 2007), and *Culex tarsalis* and *Culex pipiens* in Western North America (Reisen et al. 2004; Turell et al. 2005; Bolling et al. 2007). All of these mosquitoes commonly utilize natural wetlands and anthropogenic wetlands, including storm water ponds and structures common in densely populated urban environments.

There is also a variety of host species for WNV with varying competencies. The most competent reservoirs are frequently found in the vicinity of wetlands, but their distributions are not necessarily restricted to wetland habitat. The American Robin (*Turdus migratorius*) has been shown to be an important amplification host of WNV (Kilpatrick et al. 2006a, b). Many other host species have been evaluated for their competence in the laboratory (e.g., Komar 2003). Using a competence index calculated from laboratory data Kilpatrick et al. (2007) determined that the eight most competent hosts for WNV (which includes birds that are not obligate users of wetlands) are the Blue Jay (*Cyanocitta cristata*), Western Scrub-Jay (*Aphelocoma californica*), Black-Billed Magpie (*Pica hudsonia*), Common Grackle (*Quiscalus quiscula*), House Finch (*Carpodacus mexicanus*), House Sparrow (*Passer domesticus*) and Ring-Billed Gull (*Larus delawarensis*), and that there were numerous other avian species with moderate and low reservoir competence. Overall, there is considerable variation in vector and host competence across North America, and this variation has important consequences for WNV transmission.

Ezenwa et al. (2006, 2007) established some important links between wetland health, vector abundance, the prevalence of WNV among vector mosquitoes, and local human infection rates. These studies were undertaken in southeast Louisiana, USA, which is an area characterized by a humid subtropical climate and abundant wetlands—conditions generally considered permissive to vector-borne disease activity. Sites evaluated by Ezenwa et al. (2006, 2007) had varying intensities of wetland, forest and human developed land cover. Conventional wisdom would tend to suggest incidence of WNV in this region would be greatest around sites with high wetland cover and human population size (as indicated by developed land cover). However, this was not the case. Across these sites vector mosquito (*Culex* spp.) and human infection rates were lowest at locations with greatest wetland cover and unrelated to human development of land. Instead, infection rates among both vector mosquitoes and humans were negatively related to the richness of non-passerine host species, namely water birds.

Results from studies by Ezenwa et al. (2006, 2007) suggest that WNV incidence is strongly influenced by the diversity of the avian community, with more diverse avian communities harboring proportionally more incompetent reservoir hosts than less diverse communities. Host community composition appears important in the case of WNV, and a growing number of other zoonotic pathogens, because a greater proportion of incompetent reservoir hosts likely results in an increase in dead-end transmission events from infected mosquitoes to hosts or, conversely, a greater proportion of mosquitoes feeding on uninfected hosts. Subsequent studies on WNV incidence across the US also support this assertion (Swaddle and Calos 2008; Allan et al. 2009). Ezenwa et al. (2007) concluded that "preserving large wetland areas, and by extension, intact wetland bird communities, may represent a valuable ecosystem-based approach for controlling WNV outbreaks". Additionally, such an approach is more sustainable than management involving chemical control of vectors.

Thus knowledge of variation in vector and host competence combined with examples of mosquito and avian communities among a variety of wetlands demonstrates compelling links between wetland health and mosquito-borne disease to humans. There are important lessons to be learned from studies of WNV in North America. In particular the over-simplistic belief that wetlands promote abundance of mosquitoes and consequently exposure risk to mosquito-borne diseases does not hold when wetlands are healthy. Indeed healthy wetlands appear protective of mosquito-borne disease, by promoting rich host communities composed of many dilution hosts. While not a component of studies by Ezenwa et al. (2006, 2007), it seems likely that wetlands at sites with low vector and human infection rates would have diverse aquatic fauna and increased ecological regulation of vector mosquito larval abundance, and this may explain the lack of positive relationship between wetland area and adult vector abundance.

Synthesis and Conclusions

In this review we have emphasized natural mechanisms that regulate vector mosquito abundance in wetlands and transmission of vector-borne diseases to humans. We have evaluated abiotic and biotic determinants of vector mosquitoes within wetlands, determinants of host community structure around wetlands, and how vectors and hosts influence pathogen transmission and ultimately human disease incidence. In particular, we have focused on how wetland health can strongly influence disease transmission, using case studies (RRV, malaria and WNV) to illustrate how these mechanisms are influenced by wetland health. There is a relative paucity of studies that enabled us to draw mechanistic links between the local condition of wetlands and mosquito-borne disease incidence. Consequently some aspects of our evaluations remain speculative. In many respects the relative paucity of studies linking mechanisms between wetlands and vector-borne disease incidence is surprising given the importance of such links for human health. No doubt there are a variety of reasons for this, such as reactive management without data collection or a predisposition of people toward chemical control of mosquito populations. It may also reflect a fundamental lack of connection between aquatic ecology, wildlife and infectious disease transmission. Whatever the reasons, future multidisciplinary research that ties together these links (such as we have illustrated throughout this chapter) are likely to have significant application and scientific value. This review underscores the importance of promoting biodiversity through the maintenance of healthy wetlands and the beneficial and sustainable ecosystem services which have consequences for human health.

References

Aaskov JG, Mataika JU, Lawrence GW, Rabukawaqa V, Tucker M, Miles JA, Dalglish DA (1981) An epidemic of Ross River virus infection in Fiji, 1979. Am J Trop Med Hyg 30:1053–1059

Allan B, Langerhans R, Ryberg W, Landesman W, Griffin N, Katz R, Oberle B, Schutzenhofer M, Smyth K, de St. Maurice A, Clark L, Crooks K, Hernandez D, McLean R, Ostfeld R, Chase J (2009) Ecological correlates of risk and incidence of West Nile virus in the United States. Oecologia 158:699–708

Alonso D, Bouma MJ, Pascual M (2011) Epidemic malaria and warmer temperatures in recent decades in an East African highland. Proc R Soc B Biol Sci 278:1661–1669

Angelon K, Petranka J (2002) Chemicals of predatory mosquitofish (*Gambusia affinis*) influence selection of oviposition site by *Culex* mosquitoes. J Chem Ecol 28:797–806

Arav D, Blaustein L (2006) Effects of pool depth and risk of predation on oviposition habitat selection by temporary pool dipterans. J Med Entomol 43:493–497

Barton PS, Aberton JG (2005) Larval development and autogeny in *Ochlerotatus camptorhynchus* (Thomson) (Diptera: Culicidae) from Southern Victoria. Proc Linn Soc New South Wales 126:261–267

Batzer DP, Wissinger SA (1996) Ecology of insect communities in nontidal wetlands. Annu Rev Entomol 41:75–100

Blaustein L, Chase JM (2007) Interactions between mosquito larvae and species that share the same trophic level. Annu Rev Entomol 52:489–507

Blaustein L, Kotler B (1993) Oviposition habitat selection by the mosquito *Culiseta longiareolata*: effects of conspecifics, food and green frog tadpoles. Ecol Entomol 18:104–108

Bolling BG, Moore CG, Anderson SL, Blair CD, Beaty BJ (2007) Entomological studies along the Colorado Front Range during a period of intense West Nile virus activity. J Am Mosq Control Assoc 23:37–46

Boully L (1998) Australian wetlands: community experiences and perceptions. In: Williams W (ed) Wetlands in a dry land: understanding for management. Environment Australia, Biodiversity Group, Canberra, pp 289–298

Boyd AM, Kay BH (2001) Solving the urban puzzle of Ross River and Barmah Forest viruses. Arbovirus Res Aust 8:14–22

Boyd AM, Hall RA, Gemmell RT, Kay BH (2001) Experimental infection of Australian brushtail possums, *Trichosurus vulpecula* (Phalangeridae: Marsupialia), with Ross River and Barmah Forest viruses by use of a natural mosquito vector system. Am J Trop Med Hyg 65:777–782

Bunn SE, Davies PM (1992) Community structure of the macroinvertebrate fauna and water quality of a saline river system in south-western Australia. Hydrobiologia 248:143–160

Cale DJ, Halse SA, Walker CD (2004) Wetland monitoring in the Wheatbelt of south-west Western Australia: site descriptions, waterbird, aquatic invertebrate and groundwater data. Conserv Sci West Aust 5:20–135

Carver S (2010) Resistance of mammal asemblage structure to dryland salinity in a fragmented landscape. J R Soc West Aust 93:119–128

Carver S, Spafford H, Storey A, Weinstein P (2009a) Dryland salinity and the ecology of Ross River virus: the ecological underpinnings of the potential for transmission. Vector-Borne Zoonotic Dis 9:611–622

Carver S, Spafford H, Storey A, Weinstein P (2009b) Colonisation of ephemeral water bodies in the Wheatbelt of Western Australia by assemblages of mosquitoes (Diptera: Culicidae): role of environmental factors, habitat and disturbance. Environ Entomol 38:1585–1594

Carver S, Bestall A, Jardine A, Ostfeld RS (2009c) The influence of hosts on the ecology of arboviral transmission: potential mechanisms influencing dengue, Murray Valley encephalitis and Ross River virus in Australia. Vector-Borne and Zoonotic Dis 9:51–64

Carver S, Storey A, Spafford H, Lynas J, Chandler L, Weinstein P (2009d) Salinity as a driver of aquatic invertebrate colonisation behaviour and distribution in the wheatbelt of Western Australia. Hydrobiologia 617:75–90

Carver S, Spafford H, Storey A, Weinstein P (2010) The roles of predators, competitors and secondary salinisation in structuring mosquito (Diptera: Culicidae) assemblages in ephemeral waterbodies of the Wheatbelt of Western Australia. Environ Entomol 39:798–810

CDC (2011). Lymphatic filariasis. http://www.cdc.gov/parasites/lymphaticfilariasis/. Accessed 6 May 2011

CDC (2013) West Nile Virus. http://www.cdc.gov/westnile/index.html. Accessed 18 Nov 2013

Chase JM, Knight TM (2003) Drought-induced mosquito outbreaks in wetlands. Ecol Lett 6:1017–1024

Chesson J (1984) Effect of Notonecta (Hemiptera: Notonectidae) on mosquitoes (Diptera:Culicidae): predation or selective oviposition? Environ Entomol 13:531–538

Clements AN (1999) The biology of mosquitoes. Sensory reception and behaviour, vol 2. CABI Publishing, New York

CSIRO and the Bureau of Meteorology (2011) Climate change in Australia. http://www.climatechangeinaustralia.gov.au/futureclimate.php. Accessed 26 April 2011

Cupp EW, Hassan HK, Yue X, Oldland WK, Lilley BM, Unnasch TR (2007) West Nile virus infection in mosquitoes in the Mid-South USA, 2002–2005. J Med Entomol 44:117–125

Dale P (1993) Australian wetlands and mosquito control—contain the pest and sustain the environment? Wetlands (Australia) 12:1–12

Dale P, Knight J (2008) Wetlands and mosquitoes: a review. Wetl Ecol Manage 16:255–276

Dale PER, Knight JM (2012) Managing mosquitoes without destroying wetlands: an eastern Australian approach. Wetl Ecol Manage 20:233–242

DeGroote J, Sugumaran R, Brend S, Tucker B, Bartholomay L (2008) Landscape, demographic, entomological, and climatic associations with human disease incidence of West Nile virus in the state of Iowa, USA. Int J Health Geogr 7:19

Dobson A (2009) Climate variability, global change, immunity, and the dynamics of infectious diseases. Ecology 90:920–927

Edgerly J, McFarland M, Morgan P, Livdahl T (1998) A seasonal shift in egg-laying behaviour in response to cues of future competition in a treehole mosquito. J Anim Ecol 67:805–818

Ezenwa VO, Godsey MS, King RJ, Guptill SC (2006) Avian diversity and West Nile virus: testing associations between biodiversity and infectious disease risk. Proc R Soc B-Biol Sci 273:109–117

Ezenwa VO, Milheim LE, Coffey MF, Godsey MS, King RJ, Guptill SC (2007) Land cover variation and West Nile virus prevalence: patterns, processes, and implications for disease control. Vector-Borne Zoonotic Dis 7:173–180

Fonseca DM, Keyghobadi N, Malcolm CA, Schaffner F, Mogi M, Fleischer RC, Wilkerson RC (2004a) Outbreak of West Nile virus in North America—response. Science 306:1473–1475

Fonseca DM, Keyghobadi N, Malcolm CA, Mehmet C, Schaffner F, Mogi M, Fleischer RC, Wilkerson RC (2004b) Emerging vectors in the *Culex pipiens* complex. Science 303:1535–1538

George R, Clarke J, English P (2006) Modern and palaeogeographic trends in the salinisation of the Western Australian Wheatbelt. Proceedings of the Australian Earth Sciences Convention 2006, Melbourne. http://www.earth2006.org.au/papers/extendedpdf/George%20Richard%20-%20Modern%20and%20palaeogeographic-extended.pdf. Accessed 22 Sept 2006

Gilbert B, Srivastava DS, Kirby KR (2008) Niche partitioning at multiple scales facilitates coexistence among mosquito larvae. Oikos 117:944–950

Gingrich JB, Anderson RD, Williams GM, O'Connor L, Harkins K (2006) Stormwater ponds, constructed wetlands, and other best management practices as potential preeding sites for West Nile virus vectors in Delaware during 2004. J Am Mosq Control Assoc 22:282–291

Githeko AK, Lindsay SW, Confalonieri UE, Patz JA (2000) Climate change and vector-borne diseases: a regional analysis. Bull World Health Organ 78:1136–1147

Gratz NG (1999) Emerging and resurging vector-borne diseases. Annu Rev Entomol 44:51–75

Greenway M (2005) The role of constructed wetlands in secondary effluent treatment and water reuse in subtropical and arid Australia. Ecol Eng 25:501–509

Greenway M, Dale P, Chapman H (2003) An assessment of mosquito breeding and control in four surface flow wetlands in tropical-subtropical Australia. Water Sci Technol 48:249–256

Hales S, Weinstein P, Souares Y, Woodward A (1999) El Nino and the dynamics of vectorborne disease transmission. Environ Health Perspect 107:99–102

Halse SA, Ruprecht JK, Pinder AM (2003) Salinisation and prospects for biodiversity in rivers and wetlands of south-west Western Australia. Aust J Bot 51:673–688

Hamer GL, Kitron UD, Brawn JD, Loss SR, Ruiz MO, Goldberg TL, Walker ED (2008) *Culex pipiens* (Diptera: Culicidae): a bridge vector of West Nile virus to humans. J Med Entomol 45:125–128

Harley D, Sleigh A, Ritchie S (2001) Ross River virus transmission, infection, and disease: a cross-disciplinary review. Clin Microbiol Rev 14:909–932

Heft DE, Walton WE (2008) Effects of the El Nino Southern Oscillation (ENSO) cycle on mosquito populations in southern California. J Vector Ecol 33:17–29

Horwitz P, Finlayson CM (2011) Wetlands as settings for human health: incorporating ecosystem services and health impact assessment into water resource management. Bioscience 61:678–688

Hubalek Z, Halouzka J (1999) West Nile fever—a reemerging mosquito-borne viral disease in Europe. Emerg Infect Dis 5:643–650

Hurst TP, Brown MD, Kay BH (2004) Laboratory evaluation of the predation efficacy of native Australian fish on Culex annulirostris (Diptera: Culicidae). J Am Mosq Control Assoc 20:286–291

Jardine A, Speldewinde P, Carver S, Weinstein P (2007) Dryland salinity and Ecosystem Distress Syndrome: human health implications. EcoHealth 4:10–17

Jardine A, Lindsay MDA, Johansen CA, Cook A, Weinstein P (2008a) Impact of dryland salinity on population dynamics of vector mosquitoes (Diptera: Culicidae) of Ross River virus in inland areas of southwestern Western Australia. J Med Entomol 45:1011–1022

Jardine A, Speldewinde P, Lindsay M, Cook A, Johansen C, Weinstein P (2008b) Is there an association between dryland salinity and Ross River virus disease in southwestern Australia? EcoHealth 5: 58–68

Jones KE, Patel NG, Levy MA, Storeygard A, Balk D, Gittleman JL, Daszak P (2008) Global trends in emerging infectious diseases. Nature 451:990–993

Juliano SA (2007) Population dynamics. J Am Mosq Control Assoc 23:265–275

Juliano SA (2009) Species interactions among larval mosquitoes: context dependence across habitat gradients. Annu Rev Entomol 54:37–56

Kay BH, Boyd AM, Ryan P, Hall RA (2007) Mosquito feeding patterns and natural infection of vertebrates with Ross River and Barmah Forest viruses in Brisbane, Australia. Am J Trop Med Hyg 76:417–423

Keesing F, Holt RD, Ostfeld R (2006) Effects of species diversity on disease risk. Ecol Lett 9:485–498

Kelly-Hope LA, Purdie DM, Kay BH (2004) Ross River virus disease in Australia, 1886–1998, with analysis of risk factors associated with outbreaks. J Med Entomol 41:133–150

Kennedy J (1942) On water-finding and oviposition by captive mosquitoes. Bull Entomol Res 32:279–301

Kiflawi M, Blaustein L, Mangel M (2003) Oviposition habitat selection by the mosquito *Culiseta longiareolata* in response to risk of predation and conspecific larval density. Ecol Entomol 28:168–173

Kilpatrick AM, Kramer LD, Campbell SR, Alleyne EO, Dobson AP, Daszak P (2005) West Nile virus risk assessment and the bridge vector paradigm. Emerg Infect Dis 11:425–429

Kilpatrick AM, Kramer LD, Jones MJ, Marra PP, Daszak P (2006a) West Nile virus epidemics in North America are driven by shifts in mosquito feeding behavior. PLoS Biol 4:e82

Kilpatrick AM, Daszak P, Jones MJ, Marra PP, Kramer LD (2006b) Host heterogeneity dominates West Nile virus transmission. Proc R Soc B-Biol Sci 273:2327–2333

Kilpatrick AM, LaDeau SL, Marra PP (2007) Ecology of west nile virus transmission and its impact on birds in the western hemisphere. Auk 124:1121–1136

Kilpatrick AM, Fonseca DM, Ebel GD, Reddy MR, Kramer L.D (2010) Spatial and temporal variation in vector competence of *Culex pipiens* and *Cx. restuans* mosquitoes for West Nile virus. Am J Trop Med Hyg 83:607–613

Knight RL, Walton WE, O'Meara GF, Reisen WK, Wass R (2003) Strategies for effective mosquito control in constructed treatment wetlands. Ecol Eng 21:211–232

Komar N (2003) West Nile virus: epidemiology and ecology in North America. Adv Virus Res 61:185–234

Lafferty KD (2009) The ecology of climate change and infectious diseases. Ecology 90:888–900

Laird M (1988) The natural history of larvel mosquito habitats. Academic Press Limited, London

Landesman WJ, Allan BF, Langerhans RB, Knight TM, Chase JM (2007) Inter-annual associations between precipitation and human incidence of West Nile virus in the United States. Vector-Borne Zoonotic Dis 7:337–343

Lau C, Weinstein P, Slaney D (2012) Imported cases of Ross River virus disease in New Zealand—a travel medicine perspective. Travel Med Infect Dis 10:129–134

Lindsay M, Oliveira N, Jasinska E, Johansen C, Harrington S, Wright A. E, Smith D (1996) An outbreak of Ross River virus disease in southwestern Australia. Emerg Infect Dis 2:117–120

Lindsay MDA, Breeze AL, Harrington SA, Johansen CA, Broom AK, Gordon CJ, Maley FM, Power SL, Jardine A, Smith DW (2005) Ross River and Barmah Forest viruses in Western Australia, 2000/01–2003/04: contrasting patterns of disease activity. Arbovirus Res Aust 9:194–201

Lindsay MD, Jardine A, Johansen CA, Wright AE, Harrington SA, Weinstein P (2007) Mosquito (Diptera: Culicidae) fauna in inland areas of South West Western Australia. Aust J Entomol 46:60–64

McKenzie NL, Burbidge AH, Rolfe JK (2003) Effect of salinity on small, ground-dwelling animals in the Western Australian Wheatbelt. Aust J Bot 51:725–740

McKenzie NL, Gibson N, Keighery GJ, Rolfe JK (2004) Patterns in biodiversity of terrestrial environments in the Western Australian Wheatbelt. Records of the Western Australian Museum Supplement No. 67, pp 293–335

Merritt RW, Dadd RH, Walker ED (1992) Feeding behavior, natural food, and nutritional relationships of larval mosquitos. Annu Rev Entomol 37:349–376

Mian L, Mulla M (1986) Survival and ovipositional response of *Culex quinquefasistus* Say (Diptera: Culicidae) to sewage effluent. Bull Soc Vector Ecol 11:1944–1946

Mian LS, Lovett J, Dhillon MS (2009) Effect of effluent-treated water on mosquito development in simulated ponds at the Prado wetlands of southern California. J Am Mosq Control Assoc 25:347–355

Mills JN, Gage KL, Khan AS (2010) Potential Influence of climate change on vector-borne and zoonotic diseases: a review and proposed research plan. Environ Health Perspect: doi:10.1289/ehp.0901389

Mokany A, Shine R (2003) Oviposition site selection by mosquitoes is affected by cues from conspecific larvae and anuran tadpoles. Aust Ecol 28:33–37

Mordecai EA, Paaijmans KP, Johnson LR, Balzer C, Ben-Horin T, de Moor E, McNally A, Pawar S, Ryan SJ, Smith TC, Lafferty KD (2013) Optimal temperature for malaria transmission is dramatically lower than previously predicted. Ecol Lett 16:22–30

Muhar A, Dale PE, Thalib L, Arito E (2000) The spatial distribution of Ross River virus infections in Brisbane: Significance of residential location and relationships with vegetation types. Environ Health Prev Med 4:184–189

Munga S, Minakawa N, Zhou GF, Mushinzimana E, Barrack OOJ, Githeko AK, Yan GY (2006) Association between land cover and habitat productivity of malaria vectors in western Kenyan highlands. Am J Tropical Med Hyg 74:69–75

O'Sullivan L, Jardine A, Cook A, Weinstein P (2008) Deforestation, mosquitoes, and ancient Rome: lessons for today. BioScience 58:756–760

Olson SH, Gangnon R, Silveira GA, Patz JA (2010) Deforestation and malaria in Mancio Lima county, Brazil. Emerg Infect Dis 16:1108

Orr BK, Resh VH (1992) Influence of *Myriophyllum aquaticum* cover on Anopheles mosquito abundance, oviposition, and larval microhabitat. Oecologia 90:474–482

Ostfeld RS (2009) Climate change and the distribution and intensity of infectious diseases. Ecology 90:903–905

Ostfeld RS, Keesing F (2000a) Biodiversity and disease risk: the case of Lyme disease. Conserv Biol 14:722–728

Ostfeld RS, Keesing F (2000b) The function of biodiversity in the ecology of vector-borne zoonotic diseases. Can J Zool 78: 2061–2078

O'Sullivan L, Jardine A, Cook A, Weinstein P (2008) Deforestation, mosquitoes, and ancient Rome: lessons for today. BioScience 58:756–760

Paaijmans KP, Read AF, Thomas MB (2009) Understanding the link between malaria risk and climate. Proc Natl Acad Sci U S A 106: 13844–13849

Paaijmans KP, Blanford S, Bell AS, Blanford JI, Read AF, Thomas MB (2010) Influence of climate on malaria transmission depends on daily temperature variation. Proc Natl Acad Sci U S A 107:15135–15139

Pascual M, Ahumada JA, Chaves LF, Rodo X, Bouma M (2006) Malaria resurgence in the East African highlands: temperature trends revisited. Proc Natl Acad Sci U S A 103:5829–5834

Pascual M, Cazelles B, Bouma MJ, Chaves LF, Koelle K (2008) Shifting patterns: malaria dynamics and rainfall variability in an African highland. Proc R Soc B-Biol Sci 275:123–132

Patz JA, Confalonieri UEC (2005) Human health: ecosystem regulation of infectious diseases, conditions and trends, The millennium ecosystem assessment report

Patz JA, Daszak P, Tabor GM, Aguirre AA, Pearl M, Epstein J, Wolfe ND, Kilpatrick AM, Foufopoulos J, Molyneux D, Bradley DJ, Working Group on Land Use Change and Disease Emergence (2004) Unhealthy landscapes: policy recommendations on land use change and infectious disease emergence. Environ Health Perspect 112:1092–1098

Patz JA, Campbell-Lendrum D, Holloway T, Foley JA (2005) Impact of regional climate change on human health. Nature 438:310–317

Petersen LR, Hayes EB (2004) Westward ho? The spread of West Nile virus. N Engl J Med 351:2257–2259

Petranka JW, Fakhoury K (1991) Evidence of chemically mediated avoidance response of ovipositing insects to blue-gills and green frog tadpoles. Copeia 1991:234–239

Pinder AM, Halse SA, McRae JM, Shiel RJ (2005) Occurence of aquatic invertebrates of the wheatbelt region of Western Australia in relation to salinity. Hydrobiologia 543:1–24

Rapport DJ, Regier HA, Hutchinson TC (1985) Ecosystem behavior under stress. Am Nat 125:617–640

Reisen WK, Lothrop HD, Chiles R, Madon M, Cossen C, Woods L, Husted S, Kramer V, Edman J (2004) West Nile virus in California. Emerg Infect Dis 10.1369–1378

Reisen WK, Cayan D, Tyree M, Barker CM, Eldridge B, Dettinger M (2008) Impact of climate variation on mosquito abundance in California. J Vector Ecol 33:89–98

Reiter P (2008) Global warming and malaria: knowing the horse before hitching the cart. Malar J 7:1–9

Reiter P (2010) Nile virus in Europe: understanding the present to gauge the future. Eurosurveillance 15:19508

Rey J, Walton W, Wolfe R, Connelly C, #039, Connell S, Berg J, Sakolsky-Hoopes G, Laderman A (2012) North American wetlands and mosquito control. Int J Environ Res Public Health 9:4537–4605

Rohr JR, Dobson AP, Johnson PTJ, Kilpatrick AM, Paull SH, Raffel TR, Ruiz-Moreno D, Thomas MB (2011) Frontiers in climate change-disease research. Trends Ecol Evol. (In Press)

Russell RC (1999) Constructed wetlands and mosquitoes: health hazards and management options—an Australian perspective. Ecol Eng 12:107–124

Russell RC (2002) Ross river virus: ecology and distribution. Annu Rev Entomol 47:1–31

Russell RC (2009) Mosquito-borne disease and climate change in Australia: time for a reality check. Aust J Entomol 48:1–7

Russell RC, Cope SE, Yound AJ, Hueston L (1998) Combatting the enemy—mosquitoes and Ross River virus in a joint military exerciese in tropical Australia. Am J Trop Med Hyg 59:S307

Sabbatani S, Fiorino S, Manfredi R (2010) The emerging of the fifth malaria parasite (*Plasmodium knowlesi*). A public health concern? Braz J Infect Dis 14:299–309

Sanford MR, Chan K, Walton WE (2005) Effects of inorganic nitrogen enrichment on mosquitoes (Diptera: Culicidae) and the associated aquatic community in constructed treatment wetlands. J Med Entomol 42:766–776

Schafer ML, Lundkvist E, Landin J, Persson TZ, Lundstrom JO (2006) Influence of landscape structure on mosquitoes (Diptera: Culicidae) and dytiscids (Coleoptera: Dytiscidae) at five spatial scales in Swedish wetlands. Wetlands 26:57–68

Silberbush A, Blaustein L (2008) Oviposition habitat selection by a mosquito in response to a predator: are predator-released kairomones air-borne cues? J Vector Ecol 33:208–211

Silver J (2008) Mosquito ecology: field sampling methods, vol 3, 3 edn. Springer, New York

Sinka M, Bangs M, Manguin S, Coetzee M, Mbogo C, Hemingway J, Patil A, Temperley W, Gething P, Kabaria C, Okara R, Van Boeckel T, Godfray HC, Harbach R, Hay S (2010) The dominant Anopheles vectors of human malaria in Africa, Europe and the Middle East: occurrence data, distribution maps and bionomic precis. Parasites Vectors 3:117

Slaney D, Derraik JGB, Weinstein P (2010) Driving disease emergence: will land-use changes beat climate change to the punch? N Z Med J 123:1–3

Smithburn KC, Jacobs HR (1942) Neutralization-tests against neurotropic viruses with sera collected in Central Africa. J Immunol 44:9–23

Spencer M, Blaustein L, Schwartz SS, Cohen JE (1999) Species richness and the proportion of predatory animal species in temporary freshwater pools: relationships with habitat size and permanence. Ecol Lett 2:157–166

Spencer M, Blaustein L, Cohen JE (2002) Oviposition habitat selection by mosquitoes (*Culiseta longiareolata*) and consequences for population size. Ecology 83:669–679

Spielman A, Andreadis TG, Apperson CS, Cornel AJ, Day JF, Edman JD, Fish D, Harrington LC, Kiszewski AE, Lampman R, Lanzaro GC, Matuschka FR, Munstermann LE, Nasci RS, Norris DE, Novak RJ, Pollack RJ, Reisen WK, Reiter P, Savage HM, Tabachnick WJ, Wesson DM (2004) Outbreak of West Nile virus in North America. Science 306:1473–1473

Steinman A, Banet-Noach C, Tal S, Levi O, Simanov L, Perk S, Malkinson M, Shpigel N (2003) West Nile virus infection in crocodiles. Emerg Infect Dis 9:887–889

Sudomo M, Chayabejara S, Duong S, Hernandez L, Wu WP, Bergquist R (2010) Elimination of Lymphatic Filariasis in Southeast Asia. Advances in Parasitology 72:205–233

Swaddle JP, Calos SE (2008) Increased avian diversity is associated with lower incidence of human West Nile infection: observation of the dilution effect. Plos One 3:e2488

Taylor MJ, Hoerauf A, Bockarie M (2010) Lymphatic filariasis and onchocerciasis. Lancet 376:1175–1185

Thullen JS, Sartoris JJ, Walton WE (2002) Effects of vegetation management in constructed wetland treatment cells on water quality and mosquito production. Ecol Eng 18:441–457

Tsai TF, Popovici F, Cernescu C, Campbell GL, Nedelcu NI (1998) West Nile encephalitis epidemic in southeastern Romania. Lancet 352: 767–771

Tucker P, Gilliland J (2007) The effect of season and weather on physical activity: a systematic review. Public Health 121:909–922

Turell MJ, Dohm DJ, Sardelis MR, Guinn MLO, Andreadis TG, Blow JA (2005) An update on the potential of North American mosquitoes (Diptera: Culicidae) to transmit West Nile virus. J Med Entomol 42:57–62

van Schie C Spafford H Carver S Weinstein P (2009) The salinity tolerance of *Aedes camptorhynchus* (Diptera: Culicidae) from two regions of southwestern Australia. Aust J Entomol 48:293–299

Verhoeven JTA, Beltman B, Bobbink R, Whigham DF, Vymazal J, Greenway M, Tonderski K, Brix H, Mander Ü (2006) Constructed wetlands for wastewater treatment. Wetlands and Natural Resource Management, vol 190. Springer, Berlin, pp 69–96

Walton W, Van Dam A, Popko D 2009. Ovipositional responses of two *Culex* (Diptera: Culicidae) species to larvivorous fish. J Med Entomol 46:1338–1343

Washburn J (1995) Regulatory factors affecting larval mosquito populations in container and pool habitats: implications for biological control. J Am Mosq Control Assoc 11:279–283

Wellborn GA, Skelly DK, Werner EE (1996) Mechanisms creating community structure across a freshwater habitat gradient. Annu Rev Ecol Syst 27:337–363

Whelan P, Merianos A, Hayes G, Krause V (1997) Ross River virus transmission in Darwin, Northern Territory, Australia. Arbovirus Res Aust 7:337–345

WHO (2010) World malaria report: 2010. WHO Press, Geneva. http://www.who.int/malaria/publications/atoz/9789241564106/en/index.html. Accessed 6 May 2011

Yadouleton A, N'Guessan R, Allagbe H, Asidi A, Boko M, Osse R, Padonou G, Kinde G, Akogbeto M (2010) The impact of the expansion of urban vegetable farming on malaria transmission in major cities of Benin. Parasites & Vectors 3:118

Wetlands, Livelihoods and Human Health

Matthew P. McCartney, Lisa-Maria Rebelo and Sonali Senaratna Sellamuttu

Abstract In developing countries millions of people live a life of subsistence agriculture, mired in poverty, with limited access to basic human needs, such as food and water. Under such circumstances wetlands, through the provision of a range of direct and indirect ecosystem services, play a vital role in supporting and sustaining peoples' livelihoods and hence, their health. This chapter discusses the role of wetlands in the context of the sustainable livelihoods framework in which wetlands are viewed as an asset for the rural poor in the form of "natural capital". The framework is used to illustrate how ecosystem services, livelihoods and health are entwined and how the ecosystem services provided by wetlands can be converted to human health either directly or via other livelihood assets. It highlights the contributions that wetlands make to basic human needs and, either directly or through transformations to other forms of livelihood capital, the support they provide to livelihoods and overall well-being.

Keywords Livelihood · Ecosystem service · Well-being · Natural capital · Physical capital · Financial capital · Social capital · Development · Poverty · Floodplain · Agriculture

Introduction

As aptly illustrated by others, the inter-relationships between wetlands and human health—a key contributor to human well-being—are complex and dynamic (Horwitz and Finlayson 2011; Horwitz et al. 2012; Finlayson and Horwitz 2015). Since livelihoods are a vital determinant of well-being many wetland-health links are

M. P. McCartney (✉) · L-M. Rebelo · S Senratna Sellamuttu
International Water Management Institute, P.O. Box 4199, Vientiane, Lao PDR
e-mail: m.mccartney@cgiar.org

L-M. Rebelo
e-mail: l.rebelo@cgiar.org

S. S. Sellamuttu
e-mail: s.senaratnasellamuttu@cgiar.org

© Springer Science+Business Media Dordrecht 2015   123
C. M. Finlayson et al. (eds.), *Wetlands and Human Health,* Wetlands: Ecology,
Conservation and Management 5, DOI 10.1007/978-94-017-9609-5_7

mediated through peoples' livelihoods. Good health is essential for people to maximise livelihood opportunities. Peoples' livelihoods are undermined by poor health and conversely health is impaired by poor livelihoods.

The links between wetlands, livelihoods and health have long been recognized. In the past, in many places, wetlands were viewed as unproductive (i.e. supporting few livelihoods) and the source of disease. For example, in Europe wetlands were widely perceived as wastelands, a source of malaria, and best drained and converted to agriculture. Consequently, during the seventeenth and eighteenth centuries major drainage schemes converted thousands of square kilometres of wetlands, much of which is now highly productive farmland (Cook and Williamson 1999). As a result, throughout much of Europe there is now a range of highly modified and degraded wetland landscapes (Acreman and José 2000). For example, in the Fenlands of eastern England, human activities have transformed all but 0.1 % of the original wetlands (Mountford et al. 2002).

Over the last 30 or 40 years, as European societies have become wealthier and food security less of an issue, societal values have changed. Wetlands are now widely perceived to be highly productive ecosystems that maintain environmental quality, support biodiversity, and contribute to livelihoods directly and indirectly (MA 2005). As a result, in Europe there is now a significant effort not only to conserve what little wetland habitat remains but to enlarge it through restoration activities (Colston 2003).

In contrast to Europe, across much of Africa and Asia, high proportions of the population remain extremely poor and food security is a key priority for many. However, there has been much less transformation of wetlands. Largely unmodified wetlands remain key elements of the landscape and are a vital resource, supporting the livelihoods of many millions of people (Wood et al 2013). For these people, their health is inextricably connected—in a myriad of direct and indirect ways—to the wetlands in which they live and work. Many wetland contributions to livelihoods result in positive health outcomes. However, there are also features of wetland ecosystems that can adversely affect livelihoods and/or result in negative health outcomes. Thus the reality is that in relation to livelihoods and health, wetlands present both good and bad prospects. They are effectively a "double edged sword" and the extent to which the good outweighs the bad depends a lot on site specific factors including exactly how people interact with wetlands and, importantly, how the wetlands are managed (Horwitz and Finlayson 2011; Horwitz et al. 2012).

The contributions that wetlands make to peoples' livelihoods arise from the interaction of the ecological functions they perform with human society (Fig. 1). Wetlands are seen to provide a wide range of "ecosystem services" that benefit livelihoods and societies. Depletion of ecosystem services is widely believed to translate into fewer livelihood benefits for people and therefore lower net human well-being. However, paradoxically many people who live in Africa and Asia are "wealthy" in terms of wetland ecosystem services but are otherwise extremely poor, experiencing hard lives, very low levels of well-being and poor health (Horwitz and Finlayson 2011; Finlayson and Horwitz 2015).

Generally rural communities are poor because they are not able to effectively capture the full benefits associated with the use of natural resources, in part because

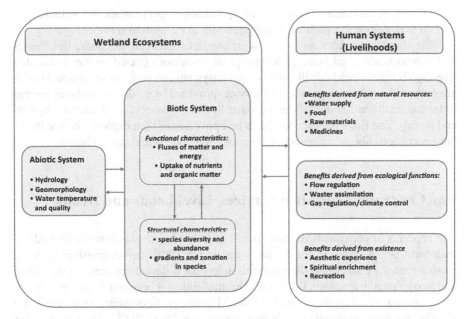

Fig. 1 A representation of the influence of wetland ecosystems on human livelihoods. (Adapted from Lorenz et al. 1997)

resources are used primarily for subsistence. To lift people out of poverty requires the identification of, and the capacity to utilize, opportunities to improve their livelihoods through economic activities: moving beyond subsistence to using the available resources in an efficient, equitable, productive and sustainable manner. Increasingly, livelihood approaches have focused on how natural resource can be used as an asset to improve peoples' well-being and promote development. This in turn inevitably requires some change to the environment. However, significantly altering wetland ecosystems for economic development can adversely affect the very ecosystem services on which the poor most depend and often results in unintended, negative consequences for the most vulnerable.

Thus there is a paradox: it is necessary to alter ecosystems to facilitate development but in altering them key ecosystem services may be undermined. The challenge for sustainable development is to obtain the right balance between socio-economic development and sustaining key ecosystem services. A prerequisite for obtaining such a balance is a comprehensive understanding of exactly how ecosystem services contribute to livelihoods and peoples' well-being and health. The dichotomy created by this paradox has been recognized by the Ramsar Convention on Wetlands. The "wise use" concept is an attempt to address it (Finlayson et al. 2011) by promoting sustainable development through the maintenance of the ecosystem services that wetlands provide. However, whilst acknowledging the importance of the wise-use concept, others argue that the "maintenance of ecological character" may reduce opportunities for the poorest and call for a more "people-centred" approach to wetland management in developing countries (Wood et al. 2013). In such circumstances, sustainable development requires that peoples' use of wetlands are related to development pathways and trade-offs in ecosystem services.

This chapter provides an overview of the links between wetlands, livelihoods and human health, and presents a conceptual framework (based on the sustainable livelihoods framework) that illustrates how ecosystem services, livelihoods and health are entwined and how the ecosystem services provided by wetlands can be converted to human health either directly or via other livelihood assets (i.e. financial, physical and social). The links between wetlands and poverty are then explored before drawing conclusions about wetlands as settings for livelihoods and human health.

The Concepts: Ecosystem Services, Livelihoods and Health

The concepts of ecosystem services, livelihoods and health are all multi-faceted human constructs. Consequently, it is not surprising that the inter-relationships between them are intricate and multi-dimensional. Below the different concepts are described and a conceptual framework illustrating the inter-linkages between them is presented.

Ecosystem services as defined by the Millennium Ecosystem Assessment are simply *"the benefits people obtain from ecosystems"* (MA 2005). The services that wetlands provide vary depending on both the biophysical characteristics of the wetland and its catchment and the presence and differing needs of beneficiaries. Four broad classes of ecosystem services have been identified (MA 2005; Fig. 2). Typically, the physical benefits from wetlands include "provisioning services" such as domestic water supply, fisheries, livestock grazing, cultivation, grass for thatching, and wild plants for food, crafts and medicinal use. Other ecosystem services are often not explicitly recognized by communities, but include a wide range of "regulating services" such as flood attenuation, maintenance of dry-season river flows,

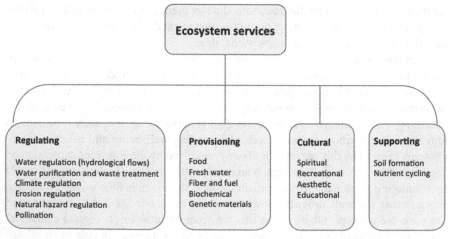

Fig. 2 Ecosystem services provided by or derived from wetlands. (McCartney et al. 2010, adapted from MA 2005)

groundwater recharge, water purification, climate regulation and erosion control, as well as a range of "supporting services" such as nutrient cycling and soil formation. In addition, people also gain nonphysical benefits from "cultural services", including spiritual enrichment, cognitive development and aesthetic experience. In many instances, different services may be closely linked. For example, where people attach spiritual value to soils and water, wetland provisioning services may be linked to cultural services. Thus, wetlands bring a wide variety of tangible and intangible benefits to large numbers of people and in this respect they provide settings for human well-being and health (Horwitz and Finlayson 2011). The way they do this is complex and multi-dimensional and is directly related to the specific features and ecological functions of the wetland.

Human Health is a key component of human well-being. Human well-being is multidimensional and defined as *the ability of people to determine and meet their needs and to have a range of choices and opportunities to fulfil their potential* (Prescott-Allen 2001). As such it requires the tackling of a diverse range of challenges—environmental, social and economic—and widening the options available to people to make a living and to participate usefully in society. The Millennium Ecosystem Assessment (MA 2005) conceptualised human well-being as representing the basic material needs for a good life, the experience of freedom, health, personal security and good social relations. Combined, these provide the conditions for physical, social, psychological and spiritual fulfilment. In this context human health is defined as *"a state of complete physical, mental, and social well-being and not merely the absence of disease or infirmity"* (WHO 2006). Many factors determine a person's health including complex interactions with the physical environment in which they live, the person's individual characteristics and behaviours, and the social and economic environment (Fig. 3).

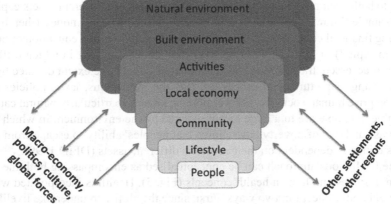

Fig. 3 The ecosystem model of settlements—determinants of health and well being. (Adapted from Barton and Grant 2006)

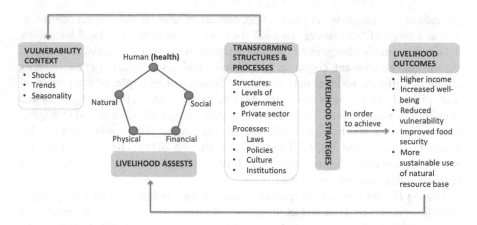

Fig. 4 Sustainable livelihoods framework as defined by UK Department for International Development (DFID 1999)

Livelihoods There are a variety of definitions but typically livelihoods are perceived as the capabilities, assets (including both material and social resources) and activities required for a means of living (Chambers and Conway 1992; Scoones 1998; DFID 1999). Livelihood strategies (i.e. the range and combination of activities and choices that people make in order to achieve desired livelihood outcomes) are influenced by the level and combination of the assets (or capitals) to which an individual has access (Fig. 4, Table 1). The sustainable livelihoods approach was developed as a way to improve understanding of the livelihoods of poor people. Again, various definitions have been proposed but most are similar to the following: *A livelihood is sustainable when it can cope with and recover from stress and shocks and can maintain or enhance its capabilities and assets, both now and in the future, while not undermining the natural resource base* (Carney et al. 1999).

Livelihood strategies can be conceived as the choices and activities individuals make to both accumulate and convert (i.e. switch) between different assets/capitals. For example, knowledge (human capital) can be used to earn money (thereby increasing financial capital) which in turn can be spent to improve education or health (human capital) or build shelter (physical capital). The manner in which different capital is accessed, transformed and accumulated is to a large extent dictated by the transforming "structures and processes" (i.e. the institutions, laws, policies etc.) that comprise human societies. Furthermore, the assets (particularly natural capital) available to people are to a large extent dictated by the environment in which they live. Past analyses of poverty have shown that peoples' ability to escape from poverty is critically dependent on their access to different assets (DFID 1999).

The livelihoods approach can be conceptualised as encompassing both the ecosystem services and human health concepts (Fig. 5). Health is incorporated within the livelihoods concept in two ways. First, since the ability to undertake livelihood strategies depends on it, health is a key component of human capital: people in

Table 1 Brief description of livelihood assets and other livelihood terminology. (Adapted from DFID 1999)

Human capital	Skills, knowledge, ability to labour and good health that together enable people to pursue different livelihood strategies and achieve livelihood objectives
Social capital	Social resources upon which people can draw in the pursuit of livelihood objectives. Typically social capital comprise networks, formal groups (i.e. operating through rules, norms and sanctions) and less formal relationships (i.e operating through trust, reciprocity and exchange)
Physical capital	Basic infrastructure (e.g. roads, shelter, water supply, electricity) and producer goods (e.g. tools and equipment) needed to support livelihoods
Financial capital	Financial resources that people use to achieve livelihood objectives
Natural capital	Natural resource stocks from which resource flows and services that are useful for livelihoods are derived
Livelihood strategy	The range and combination of activities and choices that people make/undertake to achieve their livelihood goals (including productive activities, investment strategies, reproductive choices etc.)
Livelihood outcomes	The achievements or outputs of livelihood strategies. Examples include increased income, increased well-being (including health status), improved food security etc.
Transforming structures and processes	The institutions, organisations, policies and laws that facilitate or constrain livelihood strategies and hence livelihood outcomes. They determine access to different types of capital, terms of exchanges between different types of capital and returns (financial or otherwise) to any given livelihood strategy

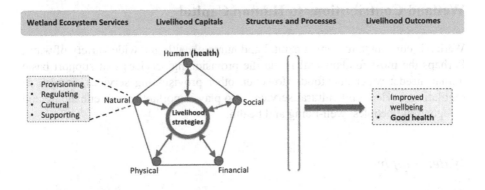

Fig. 5 A conceptual framework linking the concepts of wetland ecosystem services (natural capital), livelihoods and human health

poor health are not generally as productive as healthy people and this significantly constrains livelihood options. This is especially so for poor households in wetland dependent communities, where livelihood activities are often dependent on labour and there are fewer safety nets (Wetlands International 2010). Second, because "good health" is a fundamental constituent of human well-being, maintaining or improving health is frequently seen as a desired "outcome" (i.e. a key objective) of livelihood strategies. Thus, health is perceived as both an asset for, and an outcome of, livelihoods.

Wetland ecosystem services form an integral part of the livelihood strategy of wetland-dependent communities. The ecosystem services are, in the language of livelihoods, "natural capital". This natural capital is transformed into other livelihood assets including human capital (i.e. incorporating health) in a variety of ways. Natural capital can translate into health benefits directly. For example, through the consumption of nutritious wild foods or medicinal plants that may grow in a wetland. Natural capital can also translate into health benefits indirectly via other forms of capital. For example, wetlands often provide material (e.g. thatch and timber) that is used to build human shelters (i.e. physical capital) that contributes to human health by protecting people from the vagaries of the climate. Similarly, some wetland products may be sold and thereby converted to financial capital which in turn may be used to enhance health through the purchase of food or medicines. In some communities wetlands provide a space for community/religious activities that contribute to social cohesion thereby contributing to social capital which in turn may enhance peoples' general psychological well-being (i.e. their mental health). Thus wetlands may contribute to livelihoods and hence human health in a wide variety of ways. These interlinkages, which can be complex, are explored more fully in the sections that follow.

Wetland Contributions to Human Capital

Wetlands contribute to human capital and human health in a wide variety of ways. Perhaps the most fundamental is via the provisioning services that support basic human needs: water and food. However, other provisioning services, as well as regulating services and cultural services also play important role in enhancing and safeguarding peoples' well-being and health.

Water Supply

In locations where alternative water sources (e.g. groundwater) are scarce or dry at certain times of year, wetlands can provide water for drinking and domestic needs and also for livestock and irrigation (McCartney and Acreman 2009). In many places water abstracted directly from wetlands is essential for basic human survival. In places where water treatment facilities are unavailable, the health implications of drinking this water are entirely dependent on the natural purifying processes occurring within the wetland. These processes include sedimentation, filtration, physical and chemical immobilisation, microbial interactions and uptake by vegetation (Kadlec and Knight 1996) (Box 1). The effectiveness of these processes vary considerably between wetlands and may be highly ephemeral due to the dynamic growth and metabolic processes within them (Wetzel 2001; Prior and Jones 2002). In wetlands with short residence times during the periods of maximum sediment and pollutant loading, the benefits may not be significant (McJannet et al. 2012). Furthermore, if pollutant loadings occurring naturally or, for example, arising from upstream agriculture, exceed the physiological tolerances of key microbial and/ or plant species, degradation of the wetland will occur and effectiveness in terms of water purification is likely to decline (Gilman 1994). Effluents from industries, aquaculture ponds and domestic wastes from surrounding cities and villages may also drain into a wetland (Amarani et al. 2004), In such circumstances the health of people reliant on water from the wetland may be severely compromised.

Box 1: The filter function of the Nakivubo wetland (Source: TEEB 2010)

The Nakivubo wetland in Uganda acts as a natural filter for the effluent of the city of Kampala. Approximately 40 % of the sewage produced by the approximately 0.5 million residents of the city is discharged into the 5.5 km^2 wetland. The water from the wetland flows into Lake Victoria which is the primary source of the city's water supply. Evaluation has shown that the wastewater purification and nutrient retention services of the Nakivubo Swamp have an economic value of between US$ 1 and US$ 1.75 million.

Wild Food

Many wetlands support households through the provision of a wide range of "wild" foodstuffs, including plants, fish, birds and other animals. In the Bumwisudi wetland in Tanzania a range of wild fruits and wild vegetables are collected and eaten and in other wetlands water-birds and animals are hunted (McCartney and van Koppen 2004; Rebelo et al. 2009). Fishing, in the form of wild capture fisheries or aquaculture, is common in many wetlands, and can play a very important role in food security, not simply in terms of food provision but also in terms of vital contributions to nutrition through the provision of protein and micro-nutrients (Box 2). Wetland resources may provide the main livelihood support in periods when agricultural harvests fail completely or there is a shortfall. In the Lower Mekong Basin and many other places where people run short of food at critical times each year—so called hunger months—wetlands play a crucial role in coping strategies during periods of food shortage (Friend 2007).

Box 2: The invisible fishery of Southeast Asia

Throughout Southeast Asia, rice is the mainstay of peoples' diet. However, rice based ecosystems are not important for rice alone. They often harbor a highly diverse set of organisms that provide multiple benefits including pest control and maintenance of soil fertility as well as being an important food source in their own right (Roger et al. 1991). Some rice based ecosystems contain more than 100 species which are useful to rural communities

The vast majority of rural people in the Lower Mekong Basin harvest, consume and sell aquatic resources but only a small proportion could be considered "professional" fishers. For the majority fishing provides a way of diversifying livelihood activities and a safety net that can be relied on in the face of crop failure and other food shortages (Friend 2007).

In relation to food, the "catch" from rice fields is usually modest and only sufficient for a single day. Consequently, it often goes unnoticed in official statistics. However, this "invisible" fishery can be vitally important. In Laos, fish and other aquatic organisms, caught in rice fields and associated water channels, including amphibians, molluscs, crustaceans and insects, have been found to be vitally important for nutrition. They account for a large proportion of many people's intake of protein, micronutrients and essential fatty acids (FAO 2004).

Southeast Asian rice farmers often manage aquatic habitats and resources on their paddy land to increase the harvesting of aquatic animals which contribute to household nutrition and income. In a study to characterize the diversity of aquatic resources harvested from Farmer Managed Aquatic Systems in rice farming landscapes in Cambodia, Thailand and Vietnam. Amilhat et al. (2009) recorded farmers harvesting diverse self-recruiting species: 24 locally recognized species in Cambodia, 66 in Thailand and 17 in Vietnam. Fish accounted for the largest share by weight in all areas but frogs, snails, crustaceans and insects were also important. Amphibious species, well adapted to rice farming landscapes, dominated catches of both fish and non-fish self-recruiting species.

In some places, traditional rice fields have been cultivated sustainably for many hundreds of years. However, in recent decades as the human population has risen, pressure on rice ecosystems has also increased. Agrochemical use, sedimentation, habitat loss, destruction of fish breeding grounds and destructive fishing methods have all undermined the biodiversity of rice based systems. To feed more people increased rice production is essential. However, it should not be at the cost of the living aquatic resources on which so many people depend. Ways must be found to sustain changing human demands and simultaneously maintain the natural resources.

Agriculture

The needs of agriculture for flat, fertile land with a ready supply of water means that wetlands are often a potentially valuable agricultural resource. In arid and semiarid regions with seasonal rainfall patterns the capacity of wetlands to retain moisture for long periods, sometimes throughout the year and even during droughts, means that they are of particular importance for small-scale agriculture, both cultivation and grazing (Box 3). In Bangladesh many thousands of cattle that graze on flood-plain wetlands during the dry season are watered from perennially flooded areas known locally as beels (Anonymous 1997). Such sources are particularly important where surface water storage by means of dams or tanks is beyond the capacity of traditional rural communities. It is recognized that agricultural activities can result in a change in the ecological character of a wetland and as well as bringing benefits can pose a risk to other important ecosystem services.

Box 3: The water resource opportunities provided by dambos for small-scale farming in Zimbabwe and Malawi (Source: McCartney et al. 1997; Wood et al. 2009).

In Zimbabwe, with its savanna climate, dambos (seasonally saturated wet-lands) are estimated to occupy about 1.3 million ha. Populations have to cope with both seasonal and interannual shortages of water as a matter of course. Under such circumstances, wetland environments that retain water close to, or at the ground surface, represent a water reserve that can be used to bridge mid-season droughts and extend the length of the growing season. Conse-quently, the water resources of dambos are widely utilized as an alternative, or supplement to rain-fed agriculture. In the communal farming areas of Zimbabwe, many thousands of hectares are cultivated. Most often this takes the form of cultivation of maize, rice and vegetables in small gardens. The intensity of cultivation varies considerably, but in some communal regions an average of 30 % (actual values vary from 5–75 %) of dambo area is cultivated and in some instances this cultivation has been continuous for decades.

Although the importance of wetland agriculture, and its important contribution to livelihoods, is widely recognized, globally there is very little quantitative data on its extent. The global network of "Ramsar" sites (i.e., those wetlands designated as being of International Importance under the Ramsar Convention) currently contains 2170 sites covering more than 203 million ha. In both Africa and Asia, more than 90 % of these sites directly support human welfare in one way or another. In Africa, 68 % of them are listed as used for agriculture (including livestock), whilst the cor-responding proportion in Asia is 51 % (Table 2). Since the majority of Ramsar sites are conservation areas such figures certainly under represent the percentage of all wetlands in these regions used for agriculture.

Table 2 Wetland use in Ramsar sites of International Importance in Africa and Asia (million ha in parenthesis). Database analysed 23/1/14

Wetland use	Percentage of sites	
	Africa	Asia
Agriculture (including livestock)	68 (82)	51 (11)
Fisheries/aquaculture	58 (80)	56 (11)
Wetland products	35 (58)	29 (7)
Domestic water supply	15 (18)	10 (2)
Recreation/tourism/conservation	53 (54)	73 (13)
Total (any of the above)	91 (89)	96 (16)

In another example of dambo linked cultivation, in Malawi, in the dambos of Mpika and Kasungu, investments in upland reforestation and soil and water management in the dambos increased water available to irrigate crops during the dry months of September to December. Consequently, yields increased between 30% and 60%, while the area under cultivation increased by some 10–50%, depending on the wetland. These investments gave villagers enough to eat during the food scarce season from December to February and hence improved household nutrition (Wood 2009).

Spatial variations in soil and water properties make dambos difficult to utilize for large-scale agriculture but are exactly the features which provide opportunities for small-scale farmers. Wet patches mixed with dry soils mean working areas containing dambos as a single unit is difficult and generalized methods of large-scale farming are inappropriate. Attempts by European colonists in the first half of the twentieth century to drain dambos to produce uniform conditions resulted in rapid soil erosion, environmental deterioration and the drying out of dambos. However, at a small scale farmers in communal areas can use each part of the slope in a different way, thereby reducing the risks of crop failure. The use of dambos requires flexibility in approach because the extent of soil-moisture retention varies from year to year depending on the rainfall. In drier years sequential cropping may not be possible, while in wetter years although multiple cropping of greater diversity may be possible, waterlogging can be a problem. Indigenous farming practices that combine dry upland farming with wetland cultivation have adapted to this variability.

Medicine

The roots, leaves and bark of many wetland plants are used for medicinal purposes (Box 4). The efficacy of such remedies is likely to be highly variable but this is a direct contribution to human health. Health benefits will also accrue to individuals—not just those residing in or close to wetlands—through pharmaceuticals manufactured from wetland organisms (i.e. fungi, bacteria and algae). The medicinal qualities of these are good examples of the continued value of traditional knowledge to health care today (Horwitz and Finlayson 2011).

Box 4: Wetland medicines in Lao PDR (Source: Elkington et al. 2009).

In the Bueng Kiat Ngong wetlands, in Lao PDR more than 240 species of plants, belonging to 180 genera in 80 families of vascular plants are used by traditional healers in medical therapy. Plants used include *Tinspora crispa, Desmodium lanceolatum, Orthoisphon stamineas* and *Vitex trifolia*. At least 15 species have not been previously reported for medicinal properties, suggesting that their use maybe unique to Laos.

Disaster Risk Reduction

By disrupting livelihoods, natural disasters (e.g. floods, droughts, earthquakes) present significant hazards to people and serious consequences for their health, including in the worst cases, loss of life. The health consequences of these disasters may be immediate or may arise over the long-term, as a consequence of damage to infrastructure, increased incidence of disease and/or loss of water and food sources. The wide range of ecosystem services provided by wetlands can help mitigate the adverse impacts in both the short-term and the long-term (Horwitz et al. 2012).

In the short-term, because of their role in the hydrological cycle some wetlands may mitigate the immediate physical impacts of water-related disasters and hence enhance health by lessening peoples' exposure to physical hazards. Patterns of flow emanating from wetlands are significantly modified by hydrological processes that occur both within the wetland and through interactions in their catchment (Bullock and Acreman 2003). In some instances, wetlands regulate flows: attenuating floods and maintaining flow during dry periods and droughts (Box 5). Some coastal wetlands (e.g. mangroves and coastal marshes) may act as a form of natural coastal defence: reducing erosion, attenuating wave impacts and reducing the height of marine storm surges (Box 6). The risks from natural disasters are likely to increase in coming decades as a consequence of global and regional changes that include increasing storm intensity, accelerating sea level rise and land subsidence (Temmerman et al. 2013), as well as changes in land cover.

Wetlands that reduce risks, such as those described above, contribute to human well-being and health by saving lives and alleviating the immediate hazards. However, it should be noted that effects on flow and storm surges are a function not just of the presence/absence of a particular wetland, but also of a range of other biophysical factors, including topography, climate, soil, vegetation and geology. Consequently, the immediate hazard-reducing functions of wetlands depend to a large extent on location-specific characteristics that make it very difficult to generalise. Furthermore, many wetland processes are dynamic so that the role wetlands play may change over time: sometimes mitigating but sometimes enhancing the natural processes that cause hazards (e.g. floods). Hence, simple relationships between the areal coverage of particular wetland types within a catchment or along a coastline and the impact on flood flows/storm surges are generally not found (e.g. McCartney et al. 2013).

Box 5: River flood attenuation

Floodplain wetlands lying adjacent to river channels have been shown to attenuate floods by providing temporary upstream storage for water and reducing flow velocities. This phenomenon has been widely utilized as a means of flood control for many years. For example, controlled flooding of floodplain wetlands has long been used as a management strategy to protect the city of Lincoln in the UK (Wakelin et al. 1987) and 3800 ha of floodplain on the Charles River in the USA are estimated to save US$ 17 million in avoided flood damage each year (US Corps of Engineers 1972). Similarly, the flood protection role of the 20,000 ha That Luang wetland in Vientiane, Laos PDR has been estimated to be worth US$ 2.8 million per year (Gerrard 2004). Modelling studies on the River Cherwell in the UK comparing the flood reduction benefits of increasing the length of embankments or removing them altogether have shown that further embanking the river would increase flood flows by up to 150 % whilst restoring the floodplain to its pre-engineered configuration would reduce peak flow by between 10 and 15 % (Acreman et al. 2003).

Box 6: Coastal storm surge reduction by Mangroves

Storm surges caused by tropical cyclones are a major threat to low-lying coastal areas. Mangroves are widely perceived to reduce storm surge water levels by slowing the flow of water and reducing surface waves. Numerical simulations indicate that storm surge reductions range from 5 to 50 cm and wave height can be reduced by up to 75 % per kilometre of width of mangrove (McIvor et al. 2012). Modelling studies also indicate that the magnitude of energy absorbed depends on forest density, diameter of stems and roots, forest floor slope, bathymetry and the spectral characteristics (height, period etc.) of the incident waves and the tidal stage at which the wave enters the forest (Alongi 2008). However, there is very little statistically valid empirical evidence that mangroves have significantly reduced the human death toll of tsunamis. Studies following the 2004 Indian Ocean tsunami, indicated that the areas that suffered less were sheltered from direct exposure to the open sea (i.e. bays, lagoons, estuaries) and it was this, rather than the presence of absence of mangroves, which was the most significant determinant of damage and loss of life (Kerr et al. 2006; Chatenoux and Peduzzi 2007). Thus the role of mangroves in tsunami protection should not be overstated (Alongi 2008).

In the long-term, wetlands can provide sustenance and help people survive and recover from the impacts of natural disasters. For example, in the aftermath of a disaster, when damaged infrastructure and communication networks may make life difficult for survivors for long periods of time, the provisioning services of wetlands may supply the basic life support needs (e.g. drinking water, firewood, building materials and food). For example, the wetlands of the Mara River in Tanzania are used to grow food and are a major contributor to community coping strategies during times of extended drought (McCartney and van Koppen 2004).

Psychological Well-being

Many wetlands contribute to peoples' well-being directly but in intangible, non-physical, ways. Although there is little evidence specifically for wetlands, it is widely recognized that natural environments can contribute to spiritual enrichment, cognitive development and aesthetic experience. The sport and recreational opportunities provided by wetlands also improve lifestyles in places where livelihoods are not necessarily dependent on the wetland itself. All of these, so called "cultural services", contribute to human capital by contributing to psychological well-being (Horwitz et al 2012).

Wetland Contributions to Financial Capital

As people move away from a subsistence form of livelihood, financial capital is increasingly important (UNDP 2012). The need for money (e.g. to supply every day needs and for school fees etc.) often becomes a major driver of livelihood activities in wetlands. However, when monetary considerations come to the fore, modes of utilization are no longer influenced solely by the nature of the resources in the wetland: other factors, such as access to markets and changing demands in those markets, become significant. Furthermore, socio-economic differentiation within communities leads to substantial disparity in the utilization of the wetlands and the benefits derived from them (Box 7). The danger is that a focus on immediate short-term financial gain results in practices that are unsustainable in the long-term, ultimately undermining the ecosystem services on which livelihoods depend (Box 8).

Box 7: Social differentiation in benefits from the Kilombero Wetland, Tanzania (Source: McCartney and van Koppen 2004)

In the Kilombero wetland in Tanzania, people in the villages were traditionally fishers. However, in recent years, for a variety of reasons, there has been a significant increase in cultivation. The creation of national parks in the vicinity of the wetland has increased pressure on the wetland itself because villagers access to other areas has been curtailed. At an aggregate level, the contribution of wetland cultivation to total income is 66% of the approximately US$ 518 per household per year. However, this average masks important differences across households. Poor households receive 80% of their average annual income (US$ 230) from the wetland compared to 48 and 70% for the intermediate (US$ 414) and better-off households (US$ 910) respectively. Dryland cultivation contributes 25%, 50% and 7% of total income from cultivation to the better-off, intermediate and poor households respectively.

Box 8: Over-exploitation of wild foods in Zambia (Source: Masiyandima et al. 2004)

Chikanda, a vegetarian meatloaf made from peanuts and the boiled tubers of orchids (genera: *Disa, Satyrium, Habenaria, Brachycorythis* and *Eulophia),* is a popular delicacy in Zambia and other countries in Southern Africa. At Mabumba wetland in Zambia, the harvesting of wild orchid tubers for both household diet and for sale is a common practice. Traditionally harvesting entailed digging out the whole plant, followed by removal of the tuber and replanting the stem to allow the plant to regenerate. However, in recent years orchids have been harvested in ever increasing numbers and harvesters tend to no longer replant the stems. Consequently, orchid numbers are reported to be declining as a consequence of these unsustainable harvesting practices.

By contributing to household incomes, the ecosystem services of wetlands contribute to livelihood financial capital in a variety of ways. This includes both the sale of wetland products and through other forms of income generation, for example wetland tourism (Box 9). There is considerable evidence that higher income is linked to better health both through a direct effect on material conditions necessary for biological survival (not least the ability to purchase food and health care) and through an effect on social participation and opportunity to control life circumstances (Marmot 2002).

Box 9: Livelihood benefits of tourism in the Okavango (Source: Mbaiwa and Stronza 2010)

Approximately 120,000 tourists visit the Okavango Delta in Botswana each year. This makes it one of the primary tourist attractions in southern Africa. A recent study of the effects of tourism development on rural livelihoods in three villages in the delta found that communities have foregone traditional livelihood activities such as hunting and gathering, fishing, livestock and crop farming to participate in cash-based tourism related activities. This has included: the collection and sale of grass to safari companies for thatching lodges and camps, the production and sale of crafts, especially baskets and other wood carvings and beads to passing tourists and employment opportunities in community based natural resource management projects and safari companies in the Delta. In addition to individual benefits the communities as a whole also benefit from the sale of wildlife quotas (fees for animals hunted) and tourism fees (e.g. lodge and campsites). This income is used to fund social services and community development projects, including the installation of piped water to households and improvement of houses for the elderly and poor. As a result livelihoods have improved and, although health impacts were not evaluated explicitly, many local people reported improvements in overall well-being.

Wetland Contributions to Physical Capital

In many places wetlands contribute directly to physical capital through the provision of materials used for construction of both shelter and tools/instruments essential for livelihoods (Box 10). In many communities construction materials used for houses (e.g. grass for thatch and wooden poles) as well as agricultural tools (e.g. ploughs, axe handles, digging sticks) and household implements (e.g. bowls, baskets, food stores and water containers) have traditionally been manufactured from plants and other materials (e.g. clay) sourced from wetlands. In some instances clothes have been manufactured from wetland products. Thus wetlands can provide many of the basic necessities for supporting livelihoods and hence well-being/health.

Box 10: Wetland contributions to construction materials in Tanzania (Source: McCartney and van Koppen 2004)

In the Bahi wetland people from both the Bahi Sokoni and Chali Makulu villages collect natural grass (local name nkuruwili or kongoloare) for house thatching. From the Bumbwisudi wetland, tree species (local names—mkarati, mnazi, mdamdam, mpera, mgulabi, mzambarau and mikarafuu) are used for construction materials (i.e., for poles, furniture and thatching) and fuelwood. From the Buswahili wetland, women collect the papyrus reeds and use them for the manufacture of mats, which they sell for approximately US$ 0.5 each.

Wetland Contributions to Social Capital

Cultural Heritage

In many place wetland communities have developed customs, rituals and philoso-
phies that are synchronic with and reflective of the natural rhythms of the wetland
(Box 11). In such instances the wetland may be integral to community perspectives
of the world, create a sense of place, and are important cultural heritage. Hence,
wetlands can play an important role in enhancing the social cohesion (i.e. social
capital) of communities, which turn enhances peoples' health by contributing to
overall psychological well-being.

Box 11: The Lozi people on the Barotse floodplain

The Lozi people in western Zambia celebrate the flooding of the Zambezi
with the Kuombokav ceremony. The name means "to get out of the water
onto dry ground". Every year towards the end of the rainy season the Lozi
people make a ceremonial move to higher ground. When the Chief decides
that it's time to leave (anytime from February to May), all the people pack
their belongings into canoes and the whole tribe leaves together. The chief
in his barge with his family and a troop of traditionally dressed paddlers, in
the lead. It takes about 6 hours to cover the distance between the dry season
capital Lealui, and the wet season capital Limulunga. There the successful
move is celebrated with traditional singing and dancing. This ceremony dates
back more than 300 years when the Lozi people broke away from the great
Lunda Empire to come and settle in the upper regions of the Zambezi. The
vast plains with abundant fish was ideal for settlement but the annual floods
could not be stopped, so every year they move to higher ground until the rainy
season passes.

Institutions

Given the importance of the benefits derived from wetlands it would be surprising not to find institutions and management practices endemic to rural populations
that utilize them. In the past indigenous practices (e.g. related to land tenure within
wetlands or fishing rights) depended to a large extent on the ability of communities
to make and defend management rules. This required effective and credible local
authorities; typically traditional leaders who often derived authority from their ancestors (McCartney and Van Koppen 2004).

In recent decades, there have been radical socio-political transformations
throughout much of Africa and Asia and many of the traditional institutions governing wetland use have become less effective. Increasingly institutions formed via

formal government statutes (e.g. village committees etc.) have "officially" taken over with varying degrees of success. In many places the reality is that institutions are evolving as hybrids of modern and traditional arrangements linked together in complex and fluid networks (Cleaver and Frank 2005). Nevertheless, whatever form they take, if local people have a role to play in resource management and derive benefits from the resources around them, these institutions represent an important form of social capital, that can contribute to peoples well-being and health by: (i) by safeguarding opportunities for the poorest in communities; (ii) ensuring more equitable distribution of benefits; (iii) reducing conflict; and (iv) fostering sustainability. Under the right circumstances they enhance both conservation and rural development thereby contributing to well-being and health. The decentralization of resource management to communities has the potential to simultaneously promote conservation and development (Blaikie 2006; Mbaiwa 2005; Taylor 2002) (Box 12).

Box 12: Establishing institutions to sustainably manage Mekong wetland resources (Source: Friend 2007)

Wetland resources are fundamental to the livelihoods and health of many people living in the Mekong basin. Ensuring that these resources and the people who depend on them are adequately represented in management and decision-making processes is critical for sustainability. In Vietnam, the Mekong Wetlands Biodiversity Conservation and Sustainable use Program (MWBP) supported the establishment of Natural Resource Management Groups (NRMGs) in the Tram Chim National Park and Lang San Wetlands Reserve. Prior to these groups being established the managing authorities viewed local people as the main threat to conservation and their principal objective was to keep local people from encroaching into the park. This approach aggravated conflict between the park authorities and local people and failed to reduce pressure on resources. The NRMGs were designed to promote dialogue between local people and park authorities with the ultimate aim of enabling co-management of the wetland resources. Resource Management Plans were developed that permit access to a designated 100-ha area of the park for members of registered NRMGs, but also provide specific guidelines on what resources can be used. For example, there are restrictions on the species of fish that can be caught and the mesh size of fishing nets.

The establishment of the NRMGs has been successful in building trust between the park authorities and local people. In other places the MWBP encouraged local administrations to recognize conservation zones in seasonally flooded forests in order to protect plants important for the production of traditional medicines. In yet other places, MWBP built on traditional institutional arrangements designed to conserve fisheries through the establishment of fishery conservation zones to protect fish spawning grounds and the banning of destructive fishing practices. In many cases traditional management practices made special allowance for poorer households enabling them to use fishing gear for subsistence purposes that is otherwise banned and allowing to them to fish from community resources that are otherwise reserved for times of crisis (such as a death in the community) or a community celebration (such as a wedding).

Wetlands and Poverty

Poverty is, like livelihoods, a complex, multi-faceted phenomenon. One simple definition is "pronounced deprivation of well-being" (World Bank 2001a). Poor health is a common consequence of poverty and health is very often a key priority for poor people. For many, paying to treat health problems or losing household labour due to poor health (or the need to look after a sickly family member) can significantly undermine livelihood strategies and push households deeper into poverty. Therefore, improving nutrition and health can significantly improve livelihood productivity and reduce poverty.

For communities dependent on wetland resources, the ways in which ecosystem services integrate with other livelihood capitals are important in influencing poverty. Degradation or loss of wetland functionality often reduces the availability of food and other important services that support livelihoods. Consequently, wetland degradation and high levels of poverty often go hand-in-hand.

In a review of seven wetland case studies in Africa and Asia, the links between wetland degradation and poverty were clear (Senaratna Sellamuttu et al. 2008). However, whether poverty was a driver of wetland degradation or its result varied from case to case (Box 13). What is clear is that once wetland degradation began, a vicious spiral set in with one problem making the other worse with ever-deepening environmental degradation and poverty (Senaratna Sellamuttu et al. 2008). Modifying the ecological character of a wetland—deliberately or otherwise—can have a significant impact on livelihoods and poverty (Box 14).

Box 13: Wetland degradation and poverty linkages (Source: Senaratna Sellamuttu et al. 2008).

Poverty as a driver of wetland degradation

Lake Fundudzi which covers 144 ha is South Africa's only inland freshwater lake. Dependence on the wetland is high as the area's primary productive resource. The lake's fisheries are the main source of protein for the majority of households and its water is used to support livestock. In an attempt to improve food security, a large number of new commercial and smallholder fruit orchards and vegetable gardens were established in the catchment and cultivated both in winter and summer. Poor land use planning resulting from fragmented institutions and poor awareness meant the clearing of natural vegetation for cultivation, and housing was haphazard and began to drive excessive lake sedimentation. This was exacerbated by cultivation on steep slopes without measures for soil-erosion control. Promoting participatory wetland rehabilitation and land use planning for sustainable land use to bolster local incomes thus became a priority.

Poverty as a result of wetland degradation

The Hadejia-Nguru wetlands constitute an inland delta in northern Nigeria, located where the Hadejia and Jamaare rivers meet in the Komodugu-Yobe Basin. The basin supports a population of 18 million, 1.5 million of whom reside within the wetland. The predominance of farming, fishing, livestock-rearing and collection of wild resources indicate a high dependence on the rich wetland ecosystems. Since 1971, a series of dams have been constructed on the main tributaries to provide water for cereal irrigation. Although the yields from intensive irrigation schemes are higher per hectare than from floodplain agriculture, the total value of wetland benefits exceeds that from the irrigation: US\$ 167 ha^{-1} from the wetland compared to US\$ 29 ha^{-1} from irrigated agriculture. Since the construction of the dams and irrigation schemes, drastic changes have occurred in the wetland. The flood extent has declined from 2000 to 413 km^2. Dam design and operation have altered both the volume and timing of water flow in the basin, subjecting some parts to prolonged flooding and others to prolonged drought. The resulting wetland degradation has undermined many key livelihoods and restricted access to infrastructure and services such as credit and markets. Livelihood failures severely aggravated poverty and resulted in abandoned villages and further ecological degradation as people exploited other natural resources to cope with the loss of primary production systems.

Box 14: Kolleru Lake—changes in ecological character affect livelihoods and poverty (Source: Senaratna Sellamuttu et al. 2012)

Kolleru Lake in Andhra Pradesh is one of the largest freshwater lakes in Asia. As well as being a vital habitat for birds (189 bird species have been recorded), it has a long history of human use and conflict. Recent changes to the use of the lake demonstrate the challenges of establishing and maintaining wise use of the wetland amongst many competing demands.

During the 1990s, the Indian Government promoted intensive food production– mostly rice and aquaculture—in the lake, which had previously been used for traditional capture fisheries and as source of drinking and domestic water. The change was partly a response to food shortages in the region, but also to realize the wetland's economic potential through selling fish to other parts of India, such as Calcutta. However, much of the aquaculture was taken over by outside business interests, meaning that there was creation of some local employment but only a small proportion of the total income generated benefitted local people.

The intensive aquaculture also caused several problems. The nets and fishing traps set on the lake blocked the entrance to it. Kolleru Lake acts as a sink for storm water, and this obstruction led to serious flooding in farms surrounding the wetland. Water quality in the lake also deteriorated rapidly as consequence of pollutants from the aquaculture as well as effluents flowing into it from the intensive rice farming, in conjunction with domestic and industrial sewage, essentially eliminating the local communities' access to safe drinking water. The water, sediment and fish from the lake became contaminated with pesticides, polycyclic aromatic hydrocarbons and heavy metals (Amarani et al. 2004). Local people were compelled to purchase drinking water from traders due to the inability to use the water from the lake for drinking purposes

At the same time, conservation groups were lobbying for Kolleru Lake to be designated a wildlife sanctuary to protect its resident and migratory bird populations. After a series of legal challenges, both in favour of the designation and objections by users of the lake, part of the lake was established as a sanctuary in 1999. Local communities lost access to traditional lands through the establishment of the protected area, over which the government assumed ownership. Despite this protection, encroachment by users of the wetland resulted in continued degradation of the ecosystem, and declining bird populations. In 2005, following further legal challenges from both sides, the aquaculture ponds were cleared from the protected area. This affected the livelihoods of local communities, many of whom were extremely poor. Destruction of the aquaculture ponds reduced the occurrence of flooding but compensation for the loss of aquaculture was slow to arrive and limited. The government provided little support for new livelihoods and so many people remain entrenched in poverty.

The sustainable livelihoods approach provides a holistic framework for analyzing the objectives, scope and priorities of progress towards poverty elimination and avoids focus on individual aspects of poverty. The ecosystem services concept provides a complementary perspective that illustrates the multiple and interconnected benefits which wetlands provide. Livelihood systems interact with wetlands across a range of spatial and temporal scales, often shaping and modifying ecosystem services. Clearly, strategies for poverty alleviation should be cognisant of wetland ecosystem services and the role they play in the livelihoods of the poor. However, as discussed in the introduction to this chapter, not all drivers and constituents of poverty are addressed by simple provision of wetland ecosystem services. A wide range of options need to be considered to lift people out of poverty (Ramsar, Resolution XI.13 (2012).

Conclusion

Throughout much of the developing world near-natural wetlands represent a significant proportion of the landscape and continue to be places where many people live and derive their livelihoods. Hence, wetlands provide the settings for the livelihoods and health of these people. However, because wetlands are diverse, dynamic and multi-functional environments the way in which they interact with peoples' livelihoods, and hence influence their health, are intricate and complex. As the examples in this chapter have illustrated, the extent of livelihood dependence on wetlands and hence the demands on wetland ecosystem services are highly site specific and influenced by a myriad of biophysical, social, economic and cultural factors.

For many people, wetlands are the basis of food security and nutrition, drinking water and many other tangible and intangible benefits that impact their health. Some health benefits are derived directly from wetlands (e.g. medicinal plants) but, in common with all forms of natural capital, many others are only enabled by switching the natural capital afforded by wetlands to other forms of livelihood capital. However, wetlands do not exist to benefit people and some functions of wetlands (e.g. providing breeding habitat for mosquitoes that transmit malaria and snails that transmit schistosomiasis) can harm health and livelihoods. From a human perspective the "efficacy" of services varies considerably between wetland types, within a single wetland type and even, spatially and temporally, within a single wetland.

It is now widely recognised that economic development is necessary but not sufficient for poverty reduction. There is the risk that by degrading and undermining the productivity and sustainability of wetlands, strategies intended to increase economic benefits can in fact undermine the natural capital on which the poorest and most vulnerable depend. Support to livelihoods is not necessarily directly congruent with conservation objectives but there can be significant livelihood and health implications of getting the balance between conservation and development wrong.

References

Acreman M, Holden J (2013) How wetlands affect floods. Wetlands 33:773–786

Acreman MC, José P (2000) Wetlands In: Acreman M (ed) The hydrology of the UK: a study of change. Routledge, London, pp 204–224

Alongi DM (2008) Mangrove forests: resilience, protection from tsunamis and responses to global climate change. Estuar Coast Shelf Sci 76:1–13

Amaraneni SR, Singh S, Joshi PK (2004) Mapping the spatial distribution of air and water pollutants in Kolleru Lake, India using geographical information systems (GIS). Manage Environ Qual Int J 15(6):584–607

Amilhat E, Lorenzen K, Morales EJ, Yakupitiyage A, Little DC (2009) Fisheries production in Southeast Asian farmer managed aquatic systems (FMAS): I. Characterisation of systems. Aquaculture 296:219–226

Barton H, Grant M (2006) A health map for the local human habitat. J Royal Soc Promot Heal 126(6):252–253

Blaikie, P. 2006. Is small really beautiful?: Community Based Natural Resource Management in Malawi and Botswana. World Development 34(11) 1942–1957 doi:10.1016/j.worlddev.2005.11.023.

Bullock A, Acreman MC (2003) The role of wetlands in the hydrological cycle. Hydrol Earth Syst Sci 7:358–389

Carney D, Drinkwater M, Rusinow T, Wanmali S, Singh N (1999) Livelihood approaches compared. DFID

Chambers R, Conway G (1992) Sustainable rural livelihoods: practical concepts for the twenty-first century. IDS Discussion Paper 296. IDS, Brighton

Chatenoux B, Peduzzi P (2007) Impacts of the 2004 Indian Ocean tsunami: analysing the potential protecting role of environmental features. Nat Hazards 40:289–304

Cleaver F, Franks T (2005) How institutions elude design: river basin management and sustainable livelihoods. BCID Research Paper No. 12. Bradford Centre for International Development. University of Bradford, UK

Colston A (2003) Beyond preservation: the challenge of ecological restoration. In: Adams WM, Mulligan M (ed) Decolonising nature: strategies for nature conservation in a post-colonial era. Earthscan, London

Cook H, Williamson T (1999) Water management in the English landscape: field marsh and meadow. Edinburgh University Press, Edinburgh, pp 273

Department for International Development (DFID) (1999) Sustainable livelihoods guidance sheets. http://www.eldis.org/vfile/upload/1/document/0901/section2.pdf. Accessed 28 Apr 2014

Elkington B, Vongtakoune S, Thaimany S (2009) Medicinal plant surveys in Kiat Ngong Wetlands and its adjacent areas. Livelihoods and Landscapes Strategy, IUCN. Cited in IUCN Baseline report—Bueng Kiat Ngong wetlands, Pathoumphone District, Champassak Province, Lao PDR (p 9)

Food and Agriculture Organization (FAO) (2004) Aquatic biodiversity in rice fields. FAO, Rome Italy http://www.fao.org/rice2004/en/f-sheet/factsheet7.pdf. Accessed 28 Apr 2014

Finlayson CM, Davidson N, Pritchard D, Milton RG & Mckay H (2011) The Ramsar Convention and ecosystem-based approaches to the wise use and sustainable development of wetlands. Journal of International Wildlife Law and Policy 14:3-4, 176–198

Finlayson CM, Howitz P (2015) Wetlands as settings for human health—the benefits and the paradox. Introduction. In Finlayson CM, Howitz P (eds) Wetlands and human health. Springer, Dordrecht, pp 1–14

Friend R (2007) Securing sustainable livelihoods through wise use of wetland resources: reflections on the experience of the Mekong Wetlands biodiversity conservation and sustainable use programme (MWBP). MWBP. Vientianne, Lao PDR

Gerrard P (2004) Integrating wetland ecosystem values into urban planning: the case of that Luang Marsh, Vientiane, Lao PDR, IUCN—The World Conservation Union Asia Regional Environmental Economics Programme and WWF Lao Country Office, Vientiane

Gilman K (1994) Hydrology and wetland conservation. Wiley, Chichester, 101 p

Horwitz P, Finlayson CM (2011) Wetlands as settings for human health: incorporating ecosystem services and health impact assessment into water resource management. BioScience 6(9):678–688

Horwitz P, Finlayson CM, Weinstein P (2012) Healthy wetlands, healthy people: a review of wetlands and human health interactions. Ramsar Technical Report 6. Secretariat of the Ramsar Convention on Wetlands, Gland, Switzerland & the World Health Organization, Geneva, Switzerland

Kadlec RH, Knight RL (1996) Treatment wetlands. Lewis Publishers/CRC Press, Boca Raton

Kerr AM, Baird AH, Campbell SJ (2006) Comments on "Coastal mangrove forests mitigated tsunami" by K. Kathiresan and N. Rajendran (Estuarine, Coastal and Shelf Science, 65 (2005) 601–606). Estuar Coast Shelf Sci 67:539–541

Lorenz C, van Dijk GM, van Hattum AGM, Cofino WP (1997) Concepts in river ecology: implications for indicator development. Regul Riv Res Manage 13:501–506

MA (Millennium Ecosystem Assessment) (2005) Ecosystems and human well-being: Wetlands and water synthesis. World Resources Institute, Washington, DC

Marmot M (2002) The influence of income on health: views of an epidemiologist. Health Affairs https://sph.uth.edu/course/occupational_envhealth/bamick/RICE%20-%20Weis%20398/Marmot_income.pdf. Accessed 19 Dec 2013

Masiyandima M, McCartney MP, van Koppen B (2004) Wetland contributions to livelihoods in Zambia. FAO Netherlands Partnership Program: sustainable development and management of wetlands pp 50

Mbaiwa JE(2005) Wildlife Resource Utilization at Moremi Game Reserve and Khwai Community Area in the Okavango Delta, Botswana. *Journal of Environmental Management,* 77(2):144–156

Mbaiwa JE, Stronza AI (2010) The effects of tourism development on rural livelihoods in the Okavango Delta, Botswana. J Sustain Tour 18(5):635–656

McCartney MP, Acreman MC (2009) Wetlands and water resources. In Maltby E, Barker T (eds) The wetlands handbook. Wiley-Blackwells, pp 357–381 (ISBN: 978-0-632-05255-4)

McCartney MP, Rebelo LM, Senaratna Sellamuttu S & de Silva, S (2010) Wetlands, agriculture and poverty reduction. Colombo, Sri Lanka: International Water Management Institute (IWMI Research Report 137) pp 39, doi: 10.5337/2010.230

McCartney MP, van Koppen B (2004) Wetland contributions to livelihoods in the United Republic of Tanzania. FAO Netherlands Partnership Program: sustainable development and management of Wetlands 42 p

McCartney MP, Chigumira F, Jackson JE (1997) The water-resource opportunities provided by dambos for small-scale farming in Zimbabwe. Presented at seminar on "The management and conservation of the wetlands of Zimbabwe". Harare, Zimbabwe, 12–14 February 1997

McCartney MP, Cai X, Smakhtin V (2013) Evaluating the flow regulating functions of natural ecosystems in the Zambezi Basin. International Water Management Institute (IWMI Research Report 148), Colombo. doi:10.5337/2013.206 51pp

McIvor AL, Spencer T, Möller I, Spalding M (2012) Storm surge reduction by mangroves. Natural Coastal Protection Series: Report 2. Cambridge Coastal Research Unit Working Paper 41. University of Cambridge, The Nature Conservancy and Wetlands International

McJannet D, Wallace J, Keen R, Hawdon A, Kemer J (2012) The filtering capacity of a tropical riverine wetland: II. Sediment and nutrient balances. Hydrol Process 26:53–72

Mountford JO, McCartney MP, Manchester SJ, Wadsworth RA (2002) Wildlife habitats and their requirements within the Great Fen Project. Final CEH report to the Great Fen Project Steering Group

Prescott-Allen R (2001) The well-being of nations:a country-by-country index of quality of life and the environment. In cooporation with International Development Research Centre, IUCN, The world conservation union, international institute for environment and development, food and agricultural organization of the united nations, map maker Ltd, UNEP World conservation monitoring centre. Island Press, Washington. CABI

Prior H, Johnes PJ (2002) Regualtion of surface water quality in a Cretaceous Chalk catchment, UK: an assessment of the relative importance of instream and wetland processes. Sci Total Environ 282–283:159–174

Rebelo L-M, McCartney MP, Finlayson CM (2009) Wetlands of sub-Saharan Africa: distribution and contribution of agriculture to livelihoods. Wetl Ecol Manage 18:557–572

Roger PA, Heong KL, Teng PS (1991) Biodiversity and susatainability of wetland rice production: role and potential of microrganisms and invertebrates. In: Hawksworth DL (ed) The biodiversity of microorganisms and invertebrates: its role in sustainable agriculture. CABI

Scoones I (1998) Sustainable rural livelihoods: a framework for analysis IDS working paper 72, IDS, Brighton

Senaratna Sellamuttu S, de Silva S, Nguyen-Khoa S, Samarakoon J (2008) Good practices and lessons learned in integrating ecosystem conservation and poverty reduction objectives in wetlands. Colombo, Sri Lanka: IWMI; Wageningen, Netherlands: Wetlands International. 73 p, on behalf of, and funded by, Wetlands international's wetlands and poverty reduction project

Senaratna Sellamuttu S, de Silva S, Nagabhatla N, Finlayson M, Pattanaik C, Prasad SN (2012) The Ramsar convention's wise use concept in theory and practice: an inter-disciplinary investigation of practice in Kolleru Lake, India. J Int Wildl Law Policy [ISI] 15(03–04):228–250. do i:10.1080/13880292.2012.749138

Taylor H (2002) Insights into participation from critical management and labour process perspectives.In B. Cooke, & U. Kothari (Eds.), Participation: The new tyranny. London: Zed Books, pp. 122–138

TEEB (2010) The Economics of Ecosystems and Biodiversity: Mainstreaming the Economics of Nature: A synthesis of the approach, conclusions and recommendations of TEEB.

Temmerman S, Meire P, Bouma TJ, Herman PMJ, Ysebaert T, de Vriend HJ (2013) Ecosystem-based coastal defence in the face of global change. Nature 504:79–83

US Corps of Engineers (1972) An overview of major wetland functions and values. US Fish and Wildlife Service, FWS/OBS-84/18

Wakelin MJ, Walker TG, Wilson D (1987) Lincoln flood alleviation scheme. Proc Inst Civ Eng 82:775–776

Wetlands International (2010) Wetlands and water sanitation and hygiene (WASH)—understanding the linkages. Wetlands International, Ede

Wetzel RG (2001) Fundamental processes within natural and constructed wetland ecosystems: short-term versus long-term objectives. Water Sci Technol 44:1–8

Wood A, Dixon A, McCartney MP (2013) People-centred wetland management In: Wood A, Dixon A, McCartney MP (eds) Wetlands management and sustainable livelihoods in Africa. Routledge and Earthscan, pp 1–42

World Bank (2001) World development report. Attacking poverty: opportunity, empowerment and security. The World Bank, Washington, DC

World Health Organization (WHO) (2006) Constitution of the world health organization. WHO. http://www.who.int/governance/eb/who_constitution_en.pdf. Accessed 30 June 2011

Wetlands and Health: How do Urban Wetlands Contribute to Community Well-being?

May Carter

Abstract Much discussion relating to interactions between wetlands and people has focused on detrimental effects on health through wetland degradation and potentially toxic exposures. In recent years, however, there is greater recognition of the role wetlands play in improving the quality of human surroundings and providing cultural ecosystem services as aesthetically pleasing places for recreation, education and spiritual development. This chapter explores positive health benefits associated with the use and enhancement of urban wetlands. Potential benefits include improved physical and psychological health, increased community connection and sense of place, and those derived from community involvement in urban conservation.

To illustrate how various human health benefits may be recorded and reported, this chapter includes a case study that explores the community benefits generated through use of the Swan Canning Riverpark in Perth, Western Australia. The Riverpark consists of more than 150 conservation reserves and recreation parklands located along the banks of the Swan and Canning Rivers—a metropolitan river system that holds great spiritual, cultural and social value for the people of Perth. In 2010, the Swan River Trust began a process of parkland assessment and survey to monitor, evaluate and report on the level of community benefit derived through use of this system.

Keywords Cultural ecosystem services · Ecosystem health · Community benefit · Community engagement · Community values · Recreation · Sense of place · Conservation · Visitor impacts · Useability

Introduction—Wetlands as Healthy Places

The links between urban green spaces and human health benefits are much studied and reported (Maller et al. 2008). Urban parks and open spaces are important sites for physical activity, relaxation and social interaction and proximity to good quality

M. Carter (✉)
School of Natural Sciences, Edith Cowan University, Joondalup, WA, Australia
e-mail: mayc@upnaway.com

© Springer Science+Business Media Dordrecht 2015

C. M. Finlayson et al. (eds.), *Wetlands and Human Health*, Wetlands: Ecology, Conservation and Management 5, DOI 10.1007/978-94-017-9609-5_8

green space is a significant factor in predicting better self-reported health (Carter 2009; Pereira et al. 2013; Francis et al. 2012). However, the same level of attention has not been paid to the positive benefits associated with urban blue spaces (rivers, lakes, streams and ponds) often found as major natural features within cities or within their parklands and open spaces. Benefits associated with wetlands are often taken granted (Horwitz and Finlayson 2011) with little consideration given to the unique qualities of water and potentially positive effects on human health and well-being.

Increasing urbanisation is resulting in loss of wetlands at a rate greater than any other type of ecosystem (Ramsar Convention on Wetlands 2012). A resolution adopted by Ramsar in 2012 includes several statements that support more sustainable approaches to wetland management. These include the need to protect natural resources that sustain urban areas, recognition that access to urban green (blue) spaces can make a positive contribution to physical and mental health, and recognition that urban populations offer significant opportunity for community participation in wetland management and restoration in their local environment (Ramsar Convention on Wetlands 2012). Where development along rivers and around wetlands is increasing (such as the residential apartments along the Swan River shown in Fig. 1), potential health benefits for local residents can be optimised through opportunities to become involved in wetland care and restoration.

Fig. 1 Inner city apartment buildings overlooking wetlands and waterbird habitat along the opposite bank of the Swan River. (Image: M. Carter)

Landscape Appearance and Preference

It is often assumed that wetland ecosystems that attract visitors need to be healthy. To some extent this is true—a wetland landscape with swampy, smelly, turbid water may not attract high levels of visitation or be highly valued by a community. However, beauty is in the eye of the beholder and even a degraded wetland landscape may appear attractive and be valued by local communities (Manuel 2003).

Places with views of rivers and lakes are often cited as preferred environmental settings and are extremely popular visitor destinations as people seek landscapes and outdoor places that provide opportunities for enjoyable physical activity, relaxation and restoration, social interaction, cultural connection and spiritual enrichment, contact with nature, and escape from busy urban environments (Ibrahim and Cordes 2008; Pigram 2006). A recent study of parkland attributes and links to improved mental health found that water features, birdlife and walking paths were associated with positive perceptions of parkland quality (Francis et al. 2012).

Visitor experience of wetlands is influenced by the quality of sensory and emotive responses with the sights, smells, sounds and feel of the landscape all playing a part (Pigram 2006). The presence of wildlife can positively or negatively influence visitor experience, depending on whether resident wildlife is seen as attractive (birds, small mammals) or problematic (mosquitoes) (Horwitz and Carter 2011). Degradation of shoreline vegetation and erosion or changes to water quantity and quality may make water bodies more difficult to access and less appealing. At another level, changes to water quantity and quality may impede participation in recreational activity, particularly water-based activities such as boating, canoeing, water-skiing or swimming (Hadwen et al. 2008b).

Landscape appearance is an important factor in how wetlands are used and valued. Many authors have attempted to explain human responses to nature. Some consider that, in western countries, the general public have strong "nature-friendliness" and recognise the intrinsic value of retaining natural environments (van den Born et al. 2001). The *biophilia* and *biophobia* hypotheses (Wilson 1984, 1993; Kellert and Wilson 1993) suggest that people have an innately emotional affiliation to other living organisms (biophilia) and an evolutionary aversion to dangerous aspects of nature such as snakes and spiders (biophobia). Further to this, emotional spectra associated with nature and natural environments moves "from attraction to aversion, from awe to indifference, from peacefulness to fear-driven anxiety" (Wilson 1993) with responses influenced by culture and experience. The idea that evolution plays a role in human response to nature expands into landscape preference. It is suggested that people universally prefer open savannah-like landscapes with views of water, with links between preference for this type of landscape and evolutionary responses to environments that safeguard survival, either through provision of food and water or protection from predators. In modern times, these landscapes are simply seen as attractive and calming, promoting positive aesthetic responses and restorative health benefits (Ulrich 1986, 1993).

In general, preference is given to open waterscapes with edges that follow a natural form, and with trees and other edge vegetation as these environments are found to be appealing, restful and enjoyable (Kaplan et al. 1998). This observation is supported by the findings of a community consultation process undertaken to identify values associated with Perth's Swan and Canning rivers (Research Solutions 2007). In that study, community members preferred a mix of landscape types, with stronger preference for retention of more natural landscapes and vegetated shore lines. Preferred recreation sites were quiet natural places with few facilities.

Sites with waterscapes are often identified as favoured or favourite places. In numerous studies relating to identification of favourite places, participants almost invariably nominated a natural setting (Korpela et al. 2008). Visitors to favourite places report experiencing restorative benefits including relaxation, stress relief, regulation of emotions and feelings and reflection on personal goals (Hartig et al. 2003; Korpela and Hartig 1996; Korpela et al. 2008). It is suggested that promoting psychologically restorative experiences in nearby favourite places might be an important factor in primary healthcare (Korpela et al. 2008).

Visitor Impacts

With increased visitation come concerns about the impacts of use and the overuse of natural environments. These concerns tend to focus on two main areas: biophysical impacts such as water pollution, site deterioration, erosion, changes in ecological characteristics, and species disruption, and psycho-social impacts like crowding and recreation quality (Hadwen et al. 2008a, b). Overuse of popular areas produces concomitant impacts that result in the loss of supporting, provisioning, or regulating ecosystem services, and the cultural ecosystem services associated with visitation for pleasure may in time become substantially reduced.

Interventions to manage visitor impacts, such as controlling access or hardening water edges to reduce erosion and mitigate ecosystem degradation may also reduce (or improve) the attractiveness of particular destinations. For some visitors, evidence of ecological damage and human intervention through built environment changes may substantially reduce the experience they seek. For others, installation of visitor services and amenities may well contribute to heightened experience through ease of access or perceptions of lower risk (Pigram 2006).

While preference for safe, visually pleasing land and waterscapes that attract people is understandable, it may also be problematic. Making nature neat and tidy with natural features "arranged for human enjoyment" may be considered culturally appropriate and the "aesthetic of care" laden with good intentions of stewardship and community pride (Nassauer 2008). Such actions, however, may cause unintended harm through habitat destruction or use of herbicides and potentially "create the antithesis of ecological health". Nassauer further voices her concern that the "picturesque has been so successful in becoming popular culture that scenic landscapes are often assumed to be ecologically healthy" (Nassauer 2008).

As the need to balance wetland ecology and visitor use receives greater attention from those involved in wetland management and from those communities who use wetland systems, how to identify and integrate community benefits into conservation management practice presents some significant challenges (McInnes 2013). In the study of community values mentioned earlier (Research Solutions 2007), retaining the ecological values of the Swan and Canning rivers, most particularly protecting and enhancing water quality, was considered to be of paramount importance. Maintaining a healthy ecosystem that supported local biodiversity and recreational pursuits such as fishing and swimming received the highest community priority. How these values are being translated into management practice is discussed in the case study presented later in this chapter.

Cultural Ecosystem Services and Human Health

To achieve sustainable health, the complex links between population health and the health of urban ecosystems need to be considered (Verrinder 2007; Neller 2000). The Millennium Ecosystem Assessment (MEA) examined how changes in ecosystem services influence human health and established actions needed "to enhance the conservation and sustainable use of ecosystems and their contributions to human well-being" (MEA 2005). Within the MEA, six types of cultural services provided by ecosystems were identified: cultural diversity and identity; cultural landscapes and heritage values; spiritual services; inspiration (such as for arts and folklore); and recreation and tourism (MEA 2005).

A study involving assessment of 29 urban wetland case studies conducted in 24 countries world-wide examined awareness of planned and serendipitous (planned plus incidental) ecosystem services relating to each site (McInnes 2013). The most commonly planned cultural ecosystem services included opportunities for educational activities, picnics and outings, nature observation and tourism, knowledge and research activities, and appreciation of aesthetic and sense of place values. The frequency of serendipitous ecosystem services was higher at all sites— particularly activities relating to recreational hunting and fishing, water sports and activities, inspiration, aesthetics and sense of place values, and long term monitoring of the site (Table 1). It was concluded that with a greater number of planned ecosystem services within each site, a greater number of serendipitous activities occurred, with larger, resilient, diverse ecosystems more able to provide a range of services (McInnes 2013) and return more potential health benefits to urban communities.

Horwitz and Finlayson (2011) identified a range of health determinants that exist in wetland settings. Of most relevance here are wetlands as "settings for mental health and psychological well-being" and "places that enrich people's lives, enable them to cope, and allow them to help others". Wetlands as settings for physical activity can also be added to this list of health determinants. River promenades and paths along lake edges are popular places for walking and cycling (Volker and

Table 1 A comparison of frequency of occurrence (% of 29 sites) of planned and serendipitous cultural ecosystem services at case study sites. (Adapted from McInnes 2013)

Cultural ecosystem service	Planned	Serendipitous
Recreation and tourism		
Recreational hunting and fishing	59	83
Water sports and activities	42	62
Picnics, outings, touring	83	90
Nature observation and nature-based tourism	72	73
Spiritual and inspirational		
Inspiration	42	79
Cultural heritage	42	45
Contemporary cultural significance	45	52
Spiritual and religious values	21	24
Aesthetic and 'sense of place' values	62	97
Scientific and educational		
Educational activities and opportunities	90	100
Important knowledge systems, and importance for research	72	83
Long-term monitoring site	55	76
Major scientific study site	31	38
'Type location' for a taxon	10	14

Kistemann 2013), providing opportunity for individuals to enhance both physical health and psychological well-being.

An important link in the relationship between natural environments and self-reported levels of physical and mental health and well-being is the perceived quality, diversity and capacity of those environments to be used for relaxation, social interaction and physical activity (Carter 2009). Several factors associated with predicting higher levels of wetland visitation and recreational use were identified by Syme et al. (2001). These factors include:

- *Accessibility* is essential if a wetland environment is to be used for visitation and if people are to attach meaning to that place.
- *Ownership* can be symbolic or real and greater feelings of ownership can result in more frequent use.
- *Participation* includes the involvement of users in maintenance and future planning as those who assist in managing a place are more likely to use it.
- *Comfort* refers to how well the space around a wetland meets basic human needs such as shelter and how pleasant an environment it is to visit.
- *Security* is a prerequisite as people who feel safe and secure in an environment are more likely to visit more often.
- *Action* involves one's ability to use a wetland environment for a variety of preferred activities.

The role that urban wetlands can play in promoting better health outcomes are discussed in more detail in each of the following sections. Three aspects of wetland use and potential health benefits are highlighted:

- Places for recreation and social activity;
- Engendering a sense of place and cultural connection; and
- Engaging people in conservation activities.

Places for Recreation and Social Activity

Water-based recreation has both aesthetic and functional appeal with a distinction made between *water-dependent* activities such as sailing, fishing, swimming or water skiing and *water-enhanced* activities where the experience of walking or picnicking may be heightened by views of water (Fig. 2). Involvement in water-based outdoor recreation activities can provide substantial personal satisfaction and enjoyment (Curtis 2003; Pigram 2006).

It could be assumed that these satisfying and enjoyable experiences are more likely to found in more naturalistic settings than in much urbanised environments. This is not necessarily the case. A German study of use of Rhine river promenades

Fig. 2 A place to relax, picnic and play in riverside parkland with views across the water to Perth city. (Image: M. Carter)

in Cologne and Dusseldorf explored the health and well-being impacts associated with these urban blue spaces (Volker and Kistemann 2013). These river promenades are much developed with constructed river walls, jetties and wharves, commercial areas with shops and cafés, linear parklands and open plazas. The view of the river with its expanse of open water and parkland vegetation are often the only natural features visible. In terms of recreational use, users reported that these spaces were "lively, vital and versatile" and they experienced a "sense of freedom". The river edges were also considered to be a "favoured meeting point" that enhanced "communication between people" and created a happy atmosphere where diverse people and social groups were brought together. Perhaps most importantly, the river provided activity spaces for passive recreation—watching others, spending time in cafés, picnicking and generally relaxing—and more active pursuits such as rowing, canoeing, kayaking, sailing, walking, jogging and cycling. Many users considered these activities to contribute to a general sense of well-being and happiness.

Engendering a Sense of Place and Cultural Connection

All urban settings, whether built or natural, contribute to sense of place, with individuals' perceptions of quality and connections to local landscapes influencing potential health outcomes (Frumkin 2003). Positive relationships between people and place have the potential to produce positive physiological, psychological, social, spiritual and aesthetic effects (DeMiglio and Williams 2008) and for many people, the presence of nature plays an important role in "place-fixing" and place attachment (Beatley 2004). Conversely, negative perceptions of natural places, particularly feelings associated with apprehension and fear lower the appeal of particular areas, reduce visitation and restrict the range and type of activities undertaken (Bixler and Floyd 1997).

In the same vein, perceptions of neighbourhood quality can significantly influence self-reported health (Bowling et al. 2006; Collins et al. 2009). Manuel (2003) explored community perceptions of small neighbourhood wetlands in Nova Scotia. Despite relatively low levels of use (only approximately half the 82 people interviewed in this study described using the nearby wetlands for recreational purposes such as skating, catching frogs, hanging out or simply enjoying nature), these spaces were highly valued as part of the neighbourhood. In particular, having open space, a peaceful environment and a place for wildlife was well regarded by study participants.

From a different perspective, indigenous connections to wetlands can play a substantial role in supporting cultural identity and spiritual connections. A study of cultural values held by Nganguraku (indigenous) people in relation to the Murray River in south-eastern Australia (Mooney and Poh-Ling 2012) found strong connections to cultural identity through links to ancestors and traditional practice. In addition, the river was a place for recreation and restoration—providing freedom and escape—factors that were seen to be strongly related to better health and well-being.

The importance of cultural connection to wetlands is also articulated by the Noongar (indigenous) people of south-western Australia through their relationship with the Swan River and its waterways. At a women's meeting held as part of collaborative management of a river trails project, one elder said:

> The history of the whole Swan River, the history of any waterway, any river or any waterway that comes under Noongar country is matriarch country and it's always been that way—and the waterway has always been a symbol of women and women's birth, and that in itself has to be highlighted as our spiritual connection to the Swan River. And that doesn't only mean the Swan River that means the whole waterways. (South West Aboriginal Land and Sea Council 2011)

Involvement in the management of the Swan River waterways is an important aspect of Noongar heritage protection and caring for country. It is essential in maintaining a strong sense of place within a rapidly changing urban landscape where the river (and many sites along it) holds spiritual and cultural significance (South West Aboriginal Land and Sea Council 2011).

Engaging People in Conservation Activities

Apart from physical and psychological health benefits associated with use of open spaces (Bedimo-Rung et al. 2005; Giles-Corti et al. 2005; Sugiyama et al. 2008), visiting wetlands and parklands can engender feelings of attachment and affective (emotional) connection, which in turn, can influence positive attitudes to natural environments (Carter 2009; Dutcher et al. 2007; Williams and Patterson 2008). In addition, building social capital through civic engagement can contribute to better mental health and feelings of general health and well-being (Wood and Giles-Corti 2008).

Encouraging people to regularly visit and become actively involved in caring for local nature reserves and parklands can play an important role in health promotion and preventive health strategies. While there is little direct evaluation of the benefits of involvement in wetland conservation activities, one Australian study of people involved in a local bushland conservation project (Moore et al. 2006) found that participants reported better general health, fewer medical visits, greater satisfaction with daily activities and a stronger sense of community belonging. Evaluation of two Chicago-based prairie conservation programs (Miles et al. 1998; Miles et al. 2000) found being physically active was only one of many benefits associated with involvement. More important benefits reported by participants included spending time in nature, taking part in something meaningful, working with others, and the satisfaction of knowing they were making a positive contribution to preserving local environments.

The Swan Canning Riverpark: River Health and Community Benefits

The Swan Canning Riverpark is located in Perth, Western Australia and comprises more than 150 public foreshore reserves, with numerous associated areas of bushland, marshlands, creeks and streams all contributing to the Swan and Canning rivers system. It is jointly managed by the Swan River Trust (the Trust) and various state and local government agencies responsible for each area of parkland and open space along the rivers' banks. The River Protection Strategy for the Swan Canning Riverpark, drafted in conjunction with numerous organisations, agencies and community members, will guide how the river system is collaboratively managed (Swan River Trust 2012). The River Protection Strategy outlines a comprehensive monitoring, evaluation and reporting process and it is this process and its approach to assessing community benefit that is the focus of this case study. At the time of writing, the Swan River Trust was undergoing a process of amalgamation with the state Department of Parks and Wildlife and the final strategy was not yet approved.

The Swan Canning Riverpark

The Swan Canning Riverpark was created by the *Swan and Canning Rivers Management Act 2006* (SCRMA). This legislation recognised the importance of the rivers as a Perth icon, with the Riverpark considered to be an important natural asset and a community resource shared by local residents and visitors alike.

Despite being highly regarded by the community, historical and current uses of the Riverpark and its catchment are affecting the qualities most valued. The river system winds its way through a highly urbanised catchment and is showing signs of continuing environmental stress, including seasonal algal blooms and diminished water quality in some areas, fish kills, severe erosion and loss of riparian vegetation (Swan River Trust not dated a). Apart from ecological concerns, these environmental conditions can directly affect the way in which the river can be used. With an expanding urban population, there is growing pressure for increased access and opportunities for many different types of recreational use—from nature-based pursuits such as bird watching, to fishing, swimming, walking or cycling along river edges, to water-based sport such as paddling and rowing, to river cruising, water-skiing, speed boating and use of other forms of motorised watercraft (Swan River Trust 2012).

While demand for river-based recreational facilities continues to grow, the local community is also increasingly aware of the physical and mental health and wellbeing benefits associated with access to high-quality natural areas such as those found within the Riverpark. There is also increasing recognition of the need to protect historical and significant Aboriginal cultural sites within the Riverpark. To ensure high levels of public involvement in Riverpark protection, the Trust supports a variety of community activities including membership of the River Guardians, a network of some 1500 individuals involved in conservation, rehabilitation and

Fig. 3 Sign for collaborative urban waterways renewal project at confluence of Bickley Brook and Canning River. (Image: M. Carter)

wildlife observation projects (Swan River Trust not dated b). There are more than 40 community groups involved in conservation and restoration sites within the Riverpark itself. In addition, there are more than 200 community groups involved in catchment care, bushland and wetland conservation and restoration of sites within the broader Swan Canning catchment. In line with the Ramsar principles for planning and management of urban wetlands (Ramsar Convention on Wetlands 2012), finding a balance between the demand for increased urban development, recreational access and retention of the natural character of the river system is a collaborative effort between Riverpark land managers, associated government and non-government agencies and the Perth community (Swan River Trust 2012). The Bickley Brook Floodplain Restoration project is an example of collaboration between federal, state and local government agencies, a regional urban landcare council and a local community group (Fig. 3).

Assessment of Ecological Health and Community Benefit

Prior to 2006, the Swan River Trust had primarily operated as a statutory planning and environmental monitoring agency. There were long standing processes for measuring water quality and other ecological indicators and associated targets were reported annually. However, the Trust did not have an established monitoring and reporting program for the range of ecological, community benefit and amenity values, required to be implemented following enactment of the SCRMA in 2007. As part of adaption to the new Act, the Trust began to develop a 'State of the Rivers' monitoring and reporting framework. This involved setting new targets, indicators and monitoring programs to reflect expanded responsibilities that included evaluation of community values, as well as the river ecology (Carter 2010).

A community survey conducted for the Trust found the Swan and Canning rivers were greatly valued for ecological purposes and for community use, as well as for values related to history, culture and spirituality (Research Solutions 2007). In particular, the community considered the Riverpark to be an iconic asset and a key feature of Perth's recreational, social and cultural landscape. It was also reported that the community wished to retain maximum levels of public access to the foreshore, participate in recreational opportunities provided by the Riverpark, and to pass on a healthy Riverpark to their children and grandchildren to use and enjoy. In order to achieve this, however, the ecological integrity of the Riverpark, particularly water quality, was identified as a key aspect of Riverpark management (Research Solutions 2007).

These findings were translated into a set of Community Values (Fig. 4) that form the basis of the monitoring and reporting framework. The four value sets are: ecosystem health; sense of place; community benefit; and economic benefit. Ecosystem health was identified as the foundation on which all of the other values are built and most important to protect. Within each value set, aspects that were seen to hold greatest importance in protecting and enhancing river ecology, community benefit and amenity were identified.

The River Protection Strategy (Swan River Trust 2012), describes each value set and important aspects as follows:

- *ECOSYSTEM HEALTH: Ecosystem health is the fundamental ecological integrity that allows the Riverpark to function as a natural system. Ecosystem health is the most important value to protect. Without a healthy ecosystem, all of the other values will decline, so its protection is paramount. Ecosystem health includes protecting water quality, environmental flow, biodiversity and foreshore condition, on which the other values depend. Aspects such as the ecological and visual quality of the broader catchment, riverbanks and foreshore vegetation must also be considered.*
- *SENSE OF PLACE: Sense of place is the condition of people feeling content, healthy and safe, as a result of the rivers simply existing; offering quiet, natural spaces. Sense of place includes the connection people have with the rivers, related to their beliefs, traditions, memories and commitment to looking after it. Sense of place means different things to different people and can be the importance of a natural area for simply existing and people knowing it is there, whether they use the area or not. Therefore, protecting this value encompasses many things and extends across a broad range of people, from inside and outside of the locality, and includes people who don't actually use the Riverpark. Protecting sense of place values also involves protecting historical sites and knowledge as well as providing opportunities to practice cultural activities.*

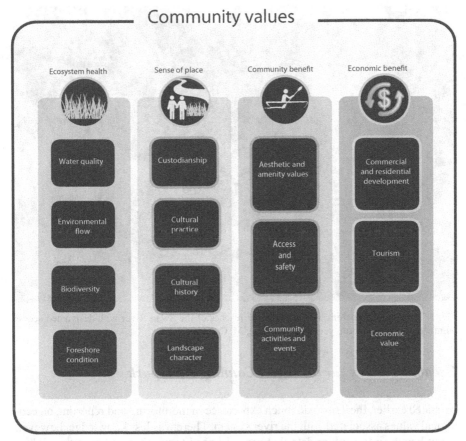

Fig. 4 Community values defined within the river protection strategy. (Source: Swan River Trust 2012)

- *COMMUNITY BENEFIT: Community benefit includes the enjoyment and comfort brought about by providing opportunities and facilities for a broad range of activities. The community benefit value incorporates the community use of the Riverpark, including aesthetics, and providing public facilities (land and water based), providing activities and events, as well as maintaining public access and safety.*
- *ECONOMIC BENEFIT: Economic benefit is the additional financial benefit of commercial and residential development and tourism opportunities gained by their proximity to the Riverpark. The Riverpark is central to the economic well-being and lifestyle of Perth's community and underpins business opportunities associated with tourism and recreational industries.*

The Swan Canning Riverpark contains numerous sites where integration of these values are evident: from boat launching points designed to minimise damage to foreshore vegetation (Fig. 5) to parklands that provide opportunities for nature discovery, recreation and relaxation, and participation in community events.

Fig. 5 Local boat launching site in the upper reaches of the Swan River with designated access points to protect foreshore vegetation. (Image: M. Carter)

Useability Index for the Swan Canning Riverpark

As stated earlier, the Trust had much experience in monitoring and reporting on ecological values associated with the river system. There was less knowledge, however, about how best to capture data that could be used to monitor and report on values associated with sense of place and community benefit. Initial review of available assessment models identified that most park management agencies measured only visitor satisfaction with facilities and services provided (Crilley et al. 2010), or simply assessed the number and purpose of visits with data most often collected through on-site surveys or observation (Ash et al. 2010; Parkin and McAlister 2010). Emerging research indicated that the personal benefits attained by visitors to a particular site (through recreational activity or contact with nature) were stronger predictors of satisfaction and positive response than the presence of infrastructure (such a pathways, toilets or car parking) or service quality (Crilley et al. 2010).

Further review of available literature suggested that rather than focusing on facilities, parks managers needed to understand and better assess potential benefits, particularly those associated with visitation setting and desired activity, and how settings might positively influence benefits attained through recreational, educational, spiritual or cultural activity. With this in mind, it was recognised by Trust staff that simply conducting visitor satisfaction surveys or collecting user information would not provide the type or quality of data required to assess whether key aspects associated with sense of place and community benefit were being addressed.

While visitor satisfaction surveys were useful in collecting specific data, a new set of indicators that could assess relevant community values was required.

At that time, a recently completed PhD study conducted in Perth (Carter 2009) had explored peoples' attitudes to natural areas and their perceptions of the quality and of green spaces in and around their neighbourhood. In this study, public green spaces included bushland, wetland and lake systems, greenways, parkland, sports ovals and playgrounds. Three key elements—diversity, useability and value—were found to be most influential in determining whether people felt that nearby green spaces made a positive contribution to their health and well-being.

Integration of quantitative and qualitative data collected in this study enabled several universal elements of "useable" green spaces to be identified. No matter the specific setting, useable green spaces needed to:

- be in good condition and look cared for;
- be accessible;
- be welcoming with clear paths and access points;
- include places where people could relax;
- include places where people could meet others;
- feel safe and comfortable;
- meet needs of multiple users; and
- be valued as part of the surrounding area.

It was decided to adapt these, and other key aspects associated with community values, to develop an index that could be used to assess the capacity of the various parkland settings within the Riverpark to contribute to community health and well-being. The final structure of the Useability Index for the Swan Canning Riverpark is illustrated in Fig. 6. Assessment items fit within two key themes: *connection* (how emotionally connected are people to this place?) and *function* (how well does this place function as an activity and/or recreation destination?). These two themes are

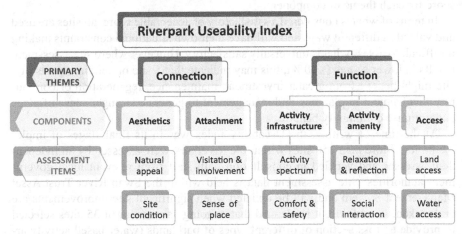

Fig. 6 Themes, components and assessment items within the Useability Index. (Source: Carter 2013)

aligned with the community values of sense of place (connection) and community benefit (function) illustrated in Fig. 4 and described above.

Within the two themes of connection and function, five components were identified.

Aesthetics and *attachment* relate to connection:

- AESTHETICS: The visual appeal of natural elements and the overall appearance of each site plays an important role in developing community connection and encouraging use
- ATTACHMENT: Engendering "a sense of place" and emotional attachment to cultural, spiritual or historical connections and landscape features plays an important role in willingness to visit, care for and protect river parklands

Activity infrastructure, activity amenity and *access* relate to function:

- ACTIVITY INFRASTRUCTURE: Appropriate activity infrastructure enables people to engage in various physical activities, recreational pastimes, social gatherings and community events
- ACTIVITY AMENITY: People seek appealing and amenable places where they can relax, reflect, meet others and socialise with family and friends
- ACCESS: Ease of access to the site (how people can get there) determines how well parklands can be used by visitors, and access within the site (such as pathways and linkages to different areas).

As illustrated in Fig. 6, ten assessment items sit under these five components. Each item includes a set of criteria that is assessed at each Riverpark site. For example, *natural appeal* includes assessment of observed water quality (whether clean, clear and odourless), attraction of cross-river views, and site appropriate trees, riparian vegetation and/or wildlife habitat. How well a site is rated (excellent to poor) in relation to each criteria determines a score (out of 10) for each assessment item. These scores can be reviewed overall (with a maximum score of 100), or by composite score for each theme or component.

In terms of what is considered a satisfactory or acceptable score, all sites are used and valued in different ways and will have their own situational constraints making it difficult to assess what is universally satisfactory. However where site assessment results in lower scores (<70%), this may indicate that a site or certain aspects of a site might benefit from greater investment, maintenance, regeneration or community involvement in planning and management to generate stronger sense of place and optimise community benefit.

At the time of writing, the Useability Index was in its final stages of implementation. More than 150 sites within the Riverpark were assessed by an independent assessor and verified in consultation with staff from relevant local government authorities. Site assessment data is held within the Swan River Trust Asset Management System and will be updated when significant site improvements are undertaken. In addition, it is planned that ongoing assessment of 35 sites selected to provide a cross-section of different types of parklands (water-based activity areas, recreation places and nature reserves) will be conducted annually by trained volunteer River Guardians. Assessment items are linked to questions included in an annual visitor satisfaction survey (conducted at the same 35 selected sites) enabling comparison of site assessment data and visitor perceptions. This information will be used to assist in prioritising investment and guide planning decisions to ensure that the health of the Swan and Canning rivers and the community who use them are maintained for future generations.

Conclusion

This chapter highlights the importance of cultural ecosystem services and their positive relationship to human health. Perhaps more importantly it demonstrates the importance of sustainable development and the need to ensure there is no further degradation or loss of urban wetlands through increasing urbanisation (Ramsar Convention on Wetlands 2012). To achieve this, the values of cultural (and other) ecosystem services associated with urban wetlands need to be clearly articulated to ensure these values inform urban planning and decision making. Local communities can greatly benefit from appropriate access to green and blue spaces, with thoughtful long term planning, monitoring and assessment designed to involve, engage and empower people to visit, value and care for urban wetlands.

References

Ash N, Blanco H, Brown C, Garcia K, Henrichs T, Lucas N, Raudsepp-Hearne C, Simpson RD, Scholes R, Tomich TP, Vira B, Zurek M (eds) (2010) Ecosystems and human well-being: a manual for assessment practitioners. Island Press, Washington, DC

Beatley T (2004). Native to nowhere: sustaining home and community in a global age. Island Press, Washington, DC

Bedimo-Rung AL, Mowen AJ, Cohen DA (2005) The significance of parks to physical activity and public health: a conceptual model. Am J Prev Med 28:159–168

Bixler RD, Floyd MF (1997) Nature is scary, disgusting, and uncomfortable. Environ Behav 29:443–467

Bowling A, Barber J, Morris R, Ebrahim S (2006) Do perceptions of neighbourhood environment influence health? Baseline findings from a British survey of aging. J Epidemiol Community Health 60:476–483

Carter M (2009) Health and the nature of urban green spaces. PhD (Environmental management), Edith Cowan University

Carter M (2010) Useability index for the Swan Canning Riverpark—Report 1: Assessment of potential to apply a useability index to monitor the contribution of the Swan Canning Riverpark to community health and well-being. Swan River Trust & Edith Cowan University, Perth

Carter M (2013) Useability index for the Swan Canning Riverpark: assessment manual. PlaceScape, Swan River Trust & Edith Cowan University, Perth

Collins PA, Hayes MV, Oliver LN (2009) Neighbourhood quality and self-rated health: a survey of eight suburban neighbourhoods in the Vancouver Census Metropolitan Area. Health Place 15:156–164

Crilley G, Weber D, Raplin R (2010) Beyond clean toilets: the importance of personal benefit attainment and service levels in park visitor satisfaction. Healthy Parks, Healthy People Congress, 10–16 April 2010 Melbourne. Parks Victoria

Curtis JA (2003) Demand for water-based leisure activity. J Environ Plan Manage 46:65–77

Demiglio L, Williams A (2008) A sense of place, a sense of well-being. In: Eyles J, Williams A (eds) Sense of place, health and quality of life. Ashgate, Hampshire

Dutcher DD, Finlay JC, Luloff AE, Johnson JB (2007) Connectivity with nature as a measure of environmental values. Environ Behav 39:473–493

Francis J, Wood L, Knuiman M, Giles-Corti B (2012) Quality or quantity? Exploring the relationship between public open space attributes and mental health in Perth, Western Australia. Soc Sci Med 74:1570–1577

Frumkin H (2003) Healthy places: exploring the evidence. Am J Public Health 93:1451–1456

Giles-Corti B, Broomhall MH, Knuiman M, Collins C, Douglas K, Ng K, Lange A, Donovan RJ (2005) Increasing walking: how important is distance to, attractiveness, and size of public open space? Am J Prev Med 28:169–176

Hadwen WL, Arthington AH, Boonington PI (2008a) Detecting visitor impacts in and around aquatic ecosystems within protected areas. Sustainable Tourism Cooperative Research Council

Hadwen WL, Hill W, Pickering C (2008b) Linking visitor impact research to visitor impact monitoring in protected areas. J Ecotourism 7:87–93

Hartig T, Evans GW, Jammer LD, Davis DS, Garling T (2003) Tracking restoration in natural and urban settings. J Environ Psychol 23:109–123

Horwitz P, Carter M (2011) Access to inland waters for tourism: ecosystem services and trade-offs. In: Crase L (ed) Water policy, tourism and recreation: lessons from Australia. Earthscan, London

Horwitz P, Finlayson MC (2011) Wetlands as settings for human health: incorpotting ecosystem services and health impact assessment into water resource management. BioScience 61:678–688

Ibrahim H, Cordes KA (2008) Outdoor recreation: enrichment for a lifetime. Sagamore, Champaign

Kaplan R, Kaplan S, Ryan RL (1998) With people in mind: design and management of everyday nature. Island Press, Washington, DC

Kellert SR, Wilson EO (eds) (1993) The biophilia hypothesis. Island Press, Washington, DC

Korpela KM, Hartig T (1996) Restorative qualities of favourite places. J Environ Psychol 16:221–233

Korpela KM, Yléna M, Tyrväinen L, Silvennoinen H. (2008) Determinants of restorative experiences in everyday favorite places. Health Place 14:636–652

Manuel PM (2003) Cultural perceptions of small urban wetlands: case studies from the Halifax Regional Municipality, Nova Scotia, Canada. Wetlands 23:921–940

Maller, Cecily, Townsend, Mardie, St Leger, Lawrence, Henderson-Wilson, Claire, Pryor, Anita, Prosser, Lauren, Moore, Megan (2008) Healthy parks, healthy people: The health benefits of contact with nature in a park context. A review of current literature (2nd ed.). Melbourne: Deakin University and Parks Victoria

McInnes RJ (2013) Recognising wetland ecosystem services within urban case studies. Mar Freshw Res 65(7):575–588

MEA (Millennium Ecosystem Assessment) (2005). Ecosystems and human well-being: synthesis. Island Press, Washington, DC

Miles I, Sullivan WC, Kuo FE (1998) Ecological restoration volunteers: the benefits of participation. Urban Ecosyst 2:27–41

Miles I, Sullivan WC, Kuo FE (2000) Psychological benefits of volunteering for restoration projects. Ecol Restor 18:218–227

Mooney C, Poh-Ling T (2012) South Australia's River Murray: social and cultural values in water planning. J Hydrol 474:29–37

Moore M, Townsend M, Oldroyd J (2006) Linking human and ecosystem health: the benefits of community involvement in conservation groups. EcoHealth J 3:255–261

Nassauer JI (2008) Cultural sustainability: aligning aesthetics and ecology. In: Carlson A, Lintott S (eds) Nature, aesthetics and environmentalism: from beauty to duty. Columbia University Press, New York

Neller AH (2000) Opportunities for bridging the gap in environmental and public health management in Australia. Ecosyst Health 6:85–91

Parkin D, Mcalister B (2010) Caring for our riverside parks and reserve: a strategy for managing riverside recreation and riparian vegetation. Australas Parks Leis 13:7–8

Pereira G, Christian H, Foster S, Boruff BJ, Bull F, Knuiman M, Giles-Corti B (2013) The association between neighborhood greenness and weight status: an observational study in Perth Western Australia. Environ Health 12:49

Pigram JJ (2006) Australia's water resources: from use to management. CSIRO Publishing, Collingwood

Ramsar Convention on Wetlands (2012) Principles for the planning and management of urban and peri-urban wetlands. In: 11th Meeting of the conference of the parties to the convention on wetlands (Ramsar I 1971) (ed). Bucharest

Research Solutions (2007) Community survey of future values and aspirations for the Swan and Canning Rivers. Swan River Trust, Perth

South West Aboriginal Land and Sea Council (2011) Final report: Swan and Canning Rivers iconic trails project. South West Aboriginal Land and Sea Council, Perth

Sugiyama T, Leslie E, Giles-Corti B, Owen N (2008) Associations of neighbourhood greenness with physical and mental health: do walking, social coherence and local social interaction explain the relationships? J Epidemiol Community Health 62:e9

Swan River Trust (2012) Draft river protection strategy. Government of Western Australia, Perth

Swan River Trust (not dated a). Issues facing the rivers. http://www.swanrivertrust.wa.gov.au/the-river-system/issues-facing-the-rivers. Accessed 2 April 2013

Swan River Trust (not dated b) River guardians. http://www.riverguardians.com/. Accessed 2 April 2013

Syme GJ, Fenton DM, Coakes S (2001) Lot size, garden satisfaction and local park and wetland visitation. Landsc Urban Plan 56:161–170

Ulrich RS (1986) Visual landscapes and psychological well-being. Landsc Urban Plan 13:29–44

Ulrich RS (1993) Biophilia, biophobia and natural landscapes. In: Kellert SR, Wilson EO (eds) The biophilia hypothesis. Island Press, Washington, DC

Van Den Born RJG, Lenders RHJ, De Groot WT, Huijsman E (2001) The new biophilia: an exploration of visions of nature in western countries. Environ Conserv 28: 65–75

Verrinder G (2007) Engaging the health sector in ecosystem viability and human health: What are barriers to, and enablers of, change? In: Horwitz P (ed) Ecology and health: people and places in a changing world. Melbourne: Organising Committee for the Asia-Pacific EcoHealth Conference 2007

Volker S, Kistemann T (2013) "I'm always entirely happy when I'm here!" Urban blue enhancing human health and well-being in Cologne and Dusseldorf, Germany. Soc Sci Med 78: 113–124

Williams DR, Patterson ME (2008) Place, leisure and well-being. In Eyles J, Williams A (eds) Sense of place, health and quality of life. Ashgate, Hampshire

Wilson EO (1984) Biophilia. Harvard University Press, Boston

Wilson EO (1993) Biophilia and the conservation ethic. In Kellert SR, Wilson EO (eds) The biophilia hypothesis. Island Press, Washington, DC

Wood L, Giles-Corti B (2008) Is there a place for social capital in the psychology of health and place? J Environ Psychol 28: 154–163

Natural Disasters, Health and Wetlands: A Pacific Small Island Developing State Perspective

Aaron P. Jenkins and Stacy Jupiter

Abstract Natural disasters in the context of public health continue to be a challenge for small island developing states (SIDS) of the Pacific. Pacific SIDS are particularly sensitive to disaster risk given geographic isolation, developing economies, lack of adaptive capacity and the interaction of climate variability with rapid environmental change. Health risks are amplified by the high levels of dependence on wetland resources and population concentration along low-lying floodplains and coastal margins. Thus, the health consequences of disasters cannot be considered in isolation from their wetland ecosystem settings. Wetlands provide protective and essential provisioning services in disasters, yet can also become vehicles for poor health outcomes. In this chapter we review the direct and indirect health consequences of interruptions to wetland ecosystem services associated with disaster events and emphasize how longer-term health effects of natural disasters can be exacerbated when wetland services are lost. We examine patterns of ill health for those populations in Pacific SIDS that are associated with wetlands and provide examples of how wetlands can either mitigate or contribute to these health outcomes. Finally, we identify opportunities and examples of improved management of wetland ecosystems for human health benefits under local to regional-scale management frameworks. Greater understanding at the interface of wetland ecology and disaster epidemiology is needed to strengthen existing models of disaster risk management and wetland conservation. We suggest applying principles of Integrated Island Management (IIM) as regionally appropriate means to guide those seeking to build this understanding.

Keywords SIDS (Small Island Developing States) · Natural disasters · River basin modification · Flooding · Cyclones · Sea level rise · Water provisioning · Food provisioning · Livelihoods · Infectious disease · Physical hazards · Psychosocial well-being · Migratory aquatic faunas · Fisheries · Disaster risk reduction · Public health · Integrated island management

A. P. Jenkins (✉)
School of Natural Sciences, Edith Cowan University, Joondalup, WA 6027, Australia
e-mail: a.jenkins@ecu.edu.au

S. Jupiter
Wildlife Conservation Society, Suva, Fiji
e-mail: sjupiter@wcs.org

© Springer Science+Business Media Dordrecht 2015
C. M. Finlayson et al. (eds.), *Wetlands and Human Health,* Wetlands: Ecology, Conservation and Management 5, DOI 10.1007/978-94-017-9609-5_9

Introduction

Natural disasters are disruptions to ecological systems that exceed people's capacity to adjust, thereby necessitating external assistance (Lechat 1976). They can be geophysical (earthquake, volcano, mass movement), meteorological (storm), hydrological (flood), climatological (extreme temperature, drought, wildfire), biological (epidemic, infestation, stampede) and extra-terrestrial (asteroid, meteorite) in nature (Below et al. 2009). Natural disasters currently affect over 200 million people annually, are frequently concentrated in and around wetland areas, and cause considerable loss of life and prolonged public health consequences (UNISDR 2005). Public health impacts of natural disasters are magnified by the interactions of urbanization, environmental degradation and climate change on floodplains, coastal margins and tectonically active areas (Kouadio et al. 2012).

Small island developing states (SIDS) in Oceania, encompassing Melanesia, Micronesia and Polynesia (Fig. 1), are particularly vulnerable to the impacts of natural disasters (Table 1). Five Pacific Island countries rank among the world's top 15 at-risk countries, including Vanuatu, Tonga, Solomon Islands, Papua New Guinea and Fiji (ADW 2012). In the context of disaster risk, vulnerability is defined as the state of susceptibility to harm from disturbance (Adger 2006) and is a function of exposure, sensitivity and adaptive capacity (IPCC 2007). Pacific SIDS have high expo-

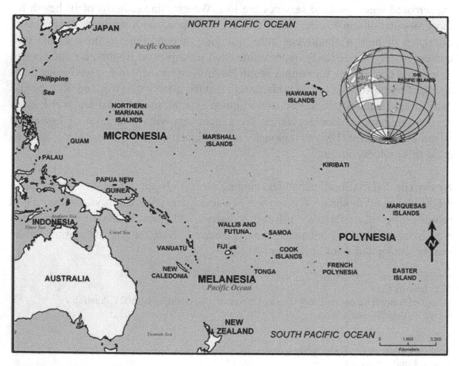

Fig. 1 Map of the Pacific Islands, showing its sub-regions and the principal island nations

Table 1 Pacific islands country estimated level of vulnerability to specific natural hazards including interacting effects. Vulnerability is depicted as high (H), medium (M) or low (L) as a function of exposure, sensitivity and adaptive capacity. Dashes indicate where the disaster type has not occurred, is unlikely to occur or cannot occur within the particular country. (Adapted from UNDP 2002)

Country	Tropical cyclones	Storm surges	Coastal floods	River floods	Drought	Earthquake	Landslide	Tsunami	Volcanic eruptions
Cook Islands	H	H	M	L	H	L	L	M	–
Federated States of Micronesia	M	M	H	H	L	L	M	–	–
Fiji	H	H	H	H	H	H	H	H	L
Kiribati	L	M	H	–	H	L	L	L	–
Republic of the Marshall Islands	H	H	H	–	H	L	L	L	–
Nauru	L	L	L	–	H	L	L	L	–
Niue	H	H	M	–	H	M	L	M	–
Palau	H	H	M	M	H	L	L	M	–
Papua New Guinea	H	H	H	H	H	H	H	H	H
Samoa	H	H	H	M	L	M	H	H	M
Solomon Islands	H	H	H	H	H	H	H	H	H
Tokelau	H	H	H	–	M	L	L	M	–
Tonga	H	H	H	L	H	H	L	H	H
Tuvalu	H	M	H	–	H	L	L	M	–
Vanuatu	H	H	H	H	H	H	H	H	H

sure to natural disasters, whose incidence and severity are on the rise (ABM and CSIRO 2011). Pacific SIDS are also particularly sensitive to disaster risk given the concentration of human populations along low-lying wetland areas (McIvor et al. 2012). Furthermore, the geographic isolation of many island communities results in low adaptive capacity, as they cannot easily access emergency services, freshwater and food following acute disturbance events (Barnett and Campbell 2010).

Following natural disasters, there are direct and indirect pathways to poorer health outcomes in wetland systems; those affected cover a broad spectrum of community members including immediate victims, rescue workers, those with lost property or livelihoods, families of the injured and those beyond the vicinity of the disaster (Galea 2007). The enormous economic costs of disasters also impact on the ability of SIDS to provide adequate recovery services: for example, annual losses from tropical cyclones and earthquakes are estimated to be as high as 6.6 % of national GDP in Vanuatu (Jha and Stanton-Geddes 2013). In this chapter we review the direct and indirect health consequences of interruptions to wetland ecosystem services associated with disaster events and emphasize how longer-term health effects of natural disasters can be exacerbated when wetland services are lost. We examine patterns of ill health for those populations in Pacific SIDS that are associated with wetlands and provide examples of how wetlands can either mitigate or contribute to these health impacts. Finally, we identify opportunities for improved management of wetland ecosystems for human health benefits under local to regional-scale management frameworks.

Impacts on Water Provisioning

It is well recognised that wetlands play a key role in the hydrological cycle, influencing both quantity and quality of available fresh water (Costanza et al. 1997; Maltby and Acreman 2011). Wetlands influence groundwater recharge, base flow maintenance, evaporation and flooding. Wetland condition also influences the quality of fresh water available for personal hydration, agriculture and industry through processes of erosion control, water purification, nutrient retention and export (Cherry 2012). Clean water on many small tropical islands is limited and vulnerable, with heightened susceptibility to contamination from inadequate sanitation and treatment facilities (Falkland 1999). The occurrence of natural disasters often results in increased contamination of water resources and disruption to water distribution and treatment facilities. Low-lying islands and atolls that rely on shallow and fragile groundwater lenses are often the most seriously affected (Dupon 1986).

Tropical storms, cyclones and associated flooding are among the most common natural disasters in the region. The Pacific region has the highest frequency of recorded cyclones, with 45 % of all cyclones reported from 1980 to 2009 (Doocy et al. 2013). Fiji alone has reported 124 natural disasters in the past 37 years, with tropical cyclones accounting for 50 % of these events (Lal et al. 2009). There is an average of five to six tropical cyclones annually in the South Pacific (Gupta 1988), resulting in major consequent impacts on freshwater provisioning and therefore acute and chronic impacts on human health. Impacts arise from direct damage to infrastructure and indirect flood-associated pollution in wetland areas from which people access water for drinking, cooking and bathing (Ellison 2009; Young et al. 2004).

For example, with wind speeds of 80–110 knots and rainfall of 300–400 mm/day, Cyclone Ami in 2003 had devastating effects on Fiji, in particular on the islands of Vanua Levu, Taveuni and the eastern islands of the Lau Group (Mosley et al. 2004). A month after the cyclone subsided, a study conducted on drinking water quality on the island of Vanua Levu showed nearly 75 % of samples did not conform to World Health Organisation (WHO) guideline values for safe drinking water (Mosley et al. 2004). This was most likely from the large amounts of silt and debris entering the water supply sources during the cyclone. Turbidity and total coliform levels had increased by 56 and 62 %, respectively, from pre-cyclone levels and poor water treatment led to this contamination being transferred through the reticulation system (Mosley et al. 2004). Communities were unaware that they were drinking water that was inadequately treated. This study also demonstrated that a simple paper strip water-quality test kit (for hydrogen sulphide, H_2S) correlated well with the lab based tests. WHO has subsequently been distributing these simple kits to remote communities to allow water supply testing post natural disaster.

The impact from storm surges on freshwater provisioning is amplified by sea level rise, which propagates storm damage further inland. While immediate damage is increased by this interaction, the compounding long-term issue of saltwater intrusion into the groundwater supply is serious (Sherif and Singh 1996). The shallow freshwater lens on atolls is the only source of fresh water other than rainwater. After a Category 5 cyclone swept a storm surge across remote Pukapuka Atoll in the Northern Cook Islands, the freshwater lens took 11 months to recover to potable conditions and remnants of the saltwater plume were still present 26 months after the saltwater incursion (Terry and Falkland 2009).

Low-lying atolls are also particularly susceptible to drought as they rely almost entirely on rainfall as a source of freshwater. In the latter part of 2011, Tuvalu declared a state of emergency after receiving less than normal rainfall for six months. Households were rationed two buckets of water a day (40 L) and the state hospital limited admissions to cope with the water rationing (NIWA 2012). The Red Cross declared that drought conditions had also caused contamination of the limited groundwater supplies (IFRC 2013). International aid agencies responded by shipping bottled water, supplying a desalination plant and increasing the number of water storage tanks in the country. This drought was attributed to La Niña, when the cooling of the surface temperature of the sea around Tuvalu leads to reduced rainfall (Salinger and Lefale 2005). A few months later, the atoll country of Tokelau also declared a state of emergency due to drought under similar circumstances of climate and geomorphological vulnerability. The Republic of the Marshall Islands, another low-lying atoll country, declared a state of disaster due to drought conditions in May 2013, with 6400 people across 15 atolls facing health, environmental, social and economic hardships due to the dry weather (IFRC 2013). Water supplies there are primarily generated by reverse osmosis units and rainwater catchments, both generally poorly maintained and limited in number (IFRC 2013).

Minimizing pollutants from reaching toxic levels in groundwater for drinking purposes is invaluable in the context of disaster recovery. Wetlands such as marshes and riparian vegetation contribute to the natural filtration of water and to the improvement of its quality (Norris 1993; Lowrance et al. 1997). The slowing of floodwaters by wetland systems allows for sediments to deposit (trapping metals and organic compounds), pollutants and nutrients to be processed, and pathogens to lose their viability or be consumed by other organisms in the ecosystem (Millennium Ecosystem Assessment 2005). High levels of nutrients in the water column, commonly associated with agricultural runoff and sewage effluent, such as phosphorus and nitrogen, can be substantially reduced or transformed by assimilation, sedimentation and other biological processes in wetlands (Dillaha et al. 1989; McKergow et al. 2003), though during periods of high flow (i.e., during heavy storms and where basins are more channeled and the gradient is steep), the extent of pollutant storage will be lower (Gaudet 1978). By contrast, losses of wetland systems can contribute to spread of waterborne bacteria. For example, a recent study in Hawaiian streams demonstrated that reductions in riparian canopy cover were associated with *Enterococcus* increases in stream water where each 1 % decrease in riparian vegetation was associated with a 4.6 % increase of *Enterococcus* (Ragosta et al. 2010).

Well-managed wetlands in disaster prone island river basins can be relied upon to mitigate some water pollution problems, however every wetland has a finite capacity to assimilate pollutants and may rapidly overload during a cyclone or flooding event (Gaudet 1978). Despite this, wetlands have a key role to play in integrated catchment-based disaster reduction and recovery strategies to address water quality issues.

Impacts on Transmission of Infectious Disease

A number of authors have predicted that climate-induced disasters will increase the incidence of infectious diseases caused by vector-borne and waterborne parasites and pathogens (Patz et al. 1996; Colwell 1996; Harvell et al. 2002; Patz et al. 2004), though this has also been challenged (Ostfeld 2009; Harper et al. 2012). Wetland alteration and other environmental damage caused by natural disasters can act as persistent drivers of infectious disease. Floods, for example, create conditions that allow mosquitoes to proliferate and increase the amount of human-mosquito contact (Gubler et al. 2001). The condition of wetlands prior to, during and after disaster events can also contribute to the collection of stagnant or slow moving water that favors mosquito breeding and associated vector-borne diseases. River-basin deforestation, river damming and rerouting all have been attributed to enhancing conditions for flooding and vector proliferation (Ahern et al. 2005). Storm surges in combination with sea level rise can also alter predominantly freshwater wetlands into increasingly brackish areas and subsequently increase breeding areas for the salt tolerant malaria vector, *Anopheles sundaicus* (Krishnamoorthy et al. 2005). In the endemic zones of the Pacific, malaria is identified as among the top five causes of non-traumatic death post-disasters (WHO 2013).

There are clear relationships between natural disasters and waterborne diseases when extremes in the hydrologic cycle cause both water shortages and floods, both of which are associated with increased diarrheal diseases (Patz et al. 2004). Within periods of water shortage, poor hygiene and the likelihood of multiple uses of the same water source (e.g., cleaning, bathing, drinking) is a major contributor to disease transmission, while other mechanisms such as concentration of pathogens may also be important (Lipp et al. 2002). The Republic of the Marshall Islands, Tokelau and Tuvalu have all experienced drought disasters since 2011, and both Tokelau and Tuvalu have experienced substantial associated diarrhea outbreaks as limited water resources became contaminated (WHO 2013).

During periods of flooding and heavy rainfall, fecal matter and associated pathogens flush from the land and contaminate drinking water sources. For example, a typhoon in Chuuk, Federated States of Micronesia, in 1971, prevented the use of the usual groundwater sources. Chuuk communities were forced to use alternative water sources, which were contaminated by pig feces, leading to an outbreak of balantidiasis (Walzer et al. 1973). In countries like Fiji, clear associations have been made with regard to increased rainfall and diarrhea cases (Singh et al. 2001) and post cyclone spikes in waterborne diseases such as typhoid have also been documented (Scobie 2011).

Damage to sanitation and sewage infrastructure associated with tropical storms and flooding also imposes serious infectious disease risk. The risks of spread of waterborne infectious diseases magnifies with lack of clean water, poor sanitation, poor nutritional status and population displacement (Dennison and Kiem 2009). Outbreaks of diarrheal illness are common after floods in low and high-income countries alike, while developing countries with poor water and sanitation infrastructure also commonly suffer from outbreaks of cholera, typhoid and other waterborne microbial diseases (Cabral 2010). In several Pacific island countries, including Fiji and Samoa, flooding events following cyclones and prolonged rainfall have been linked to outbreaks of several waterborne bacterial diseases (e.g. leptospirosis, shigellosis, typhoid), resulting in costly disaster response measures (Jenkins 2010). The probability of infectious disease increases following tropical cyclone events in proportion to disruption of public health services and the health-care infrastructure, damage to water and sanitation networks, changes in population density (especially in crowded shelters), population displacement and migration, increased environmental exposure due to damage to dwellings, and ecological changes (Shultz et al. 2005).

As people and animals are driven together in dry areas, increased contact with rodents and livestock often results in outbreaks of the bacterial infection leptospirosis (Gaynor et al. 2007; Lau et al. 2010; 2012). Leptospirosis, a bacterial disease, can be transmitted by direct contact with contaminated water. Rodents, in particular, shed large amounts of the pathogenic bacteria in their urine, and transmission occurs through contact of the skin and mucous membranes with water, damp soil or vegetation (such as sugar cane), or mud contaminated with rodent urine (Watson et al. 2007). Flooding facilitates spread of bacteria because of the proliferation of rodents and the proximity of rodents to humans on shared high ground. In 2012, Fiji

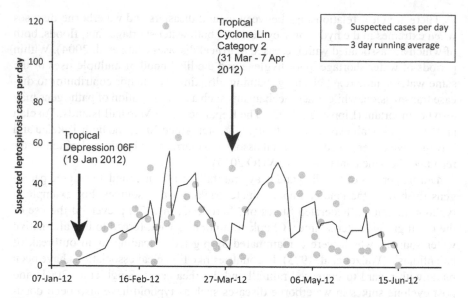

Fig. 2 Suspected leptospirosis cases after sequential flooding disasters following tropical storm events—Western Division, Fiji, 2012. (Reprinted with permission from WHO 2013)

was impacted by flooding from sequential tropical depressions that caused widespread flooding impacting much of Western Viti Levu in January 2012 and again in March 2012. These events resulted in substantial population displacement and significant post-disaster increases in leptospirosis cases three to eight weeks following each of the major flooding episodes (Fig. 2). Conservative estimates place the number of cases at 300 and the number of deaths at 25 (WHO 2013).

In the context of natural disasters such as tropical cyclones and large storms, river basin modification via logging, mining and agriculture can exacerbate the environmental exposure of communities to infectious disease (Patz 2000; Patz et al. 2004; Horwitz et al. 2012; Myers et al. 2013). The combination of increases in the severity of severe storms and rainfall and increased landscape modification is already having a notable impact on disease transmission. Around 3 % of forests are lost each year with serious consequences both on land and in adjacent water drainages (Hansen et al. 2010). Change in the diversity and abundance of species, soil dynamics, water chemistry, hydrological cycles and new forest fringe habitats creates new disease exposure dynamics (Myers and Patz 2009). Construction of dams and irrigation systems in river basins also contribute to disease emergence. Dams and irrigation have been associated with rises in schistosomiasis (Malek 1975), Rift Valley fever, filariasis, leishmaniasis, dracunculosis, onchocerciasis, and Japanese encephalitis (Harbin et al. 1993; Jobin 1999). Road building, commonly associated with logging enterprises in Pacific SIDS, has also been linked to increased incidence of dengue fever (Mackenzie et al. 2004) and diarrheal disease (Eisenberg et al. 2006).

In the Pacific Islands context, this increased environmental exposure results from interacting processes of flooding and sea level rise as river basin modification acts in concert with a changing global climate (Nicholls et al. 2007). Flooding and erosion rates are often accelerated by forest clearance (e.g. Likens et al. 1970; Costa et al. 2003). Deforestation within river basins is a major cause of flooding and landslide activity during periods of high rainfall and major storms and can prime the basin for future floods through increased sediment deposition (Cockburn et al. 1999). Where land is cleared, grazed or tilled, changes in compaction, infiltration and vegetative cover may lead to increased soil erosion and runoff (Pimentel et al. 1993; Roth 2004). A comprehensive 56 country study of forest cover and flood risk demonstrated that a 10 % loss of natural forest cover can result in a 4–28 % increase in flood frequency and a 4–8 % increase in flood duration (Bradshaw et al. 2007). The Western Pacific is also experiencing significant sea level rise (Church et al. 2006), and this interaction of river flooding and sea level rise can produce substantial increases in flood risk to populations and infrastructure. Overall, the small islands of the Pacific Ocean are among the most vulnerable to flooding (Nicholls et al. 1999) precisely because of this interaction between deforestation, wetland ecosystem degradation and climate impacts such as sea level rise and increased frequency of tropical storms (Knutson et al. 2010). Fiji, for example, has experienced an increased frequency and magnitude of flooding over the past few decades, attributable to the aforementioned combination of factors and particularly affecting populations living in river deltas and on flood plains (Lata and Nunn 2012).

Impacts on Food Provisioning and Livelihoods

Wetland ecosystems and the resources they provide are central to the food provisioning and livelihoods of Pacific island peoples. These food provisioning services can be severely impacted by various natural disasters. Health issues associated with malnutrition in the wake of disasters occur through reduced caloric, protein or micronutrient intake or ingesting toxic levels of trace elements (Cook et al. 2008). Incidence of malnutrition in Pacific SIDS is predicted to become more prolonged and widespread due to impacts associated with climate change and associated climate-induced disasters (Barnett 2007). Impaired nutritional intake is also a risk factor for mortality from infectious diseases, such as gastroenteritis and measles, which are often also more common in the post-disaster phase (Cook et al. 2008).

Natural disasters, such as tropical cyclones and severe storms, cause significant loss in agricultural production each year in the Pacific. More than 80 % of the population of the Pacific islands is rural and about 67 % depend on agriculture for their livelihoods (ADB and IFPRI 2009). Crop production for subsistence and commercial use is heavily dependent on wetlands, often located on fertile floodplain areas. Cyclone Ami, for example, caused over US$ 35 million in lost crops in Fiji in 2003 (Mackenzie et al. 2005). In Tuvalu saltwater intrusion from storm surge affected communal crop gardens on six of Tuvalu's eight islands and destroyed 60 % of

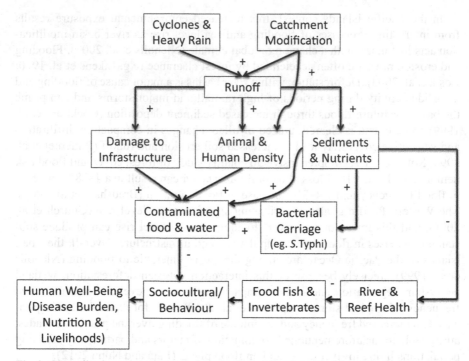

Fig. 3 Generalized conceptual model of mechanisms determining health and livelihood outcomes during cyclones or heavy rainfall in the context of high island river basins. + and − indicate positive and negative contribution or affect on the following factor and *dotted lines* show the mediating effect of socio-cultural behaviour

traditional pit gardens (ADB and IFPRI 2009). High risk of flooding in river catchments also threatens food production. Heavy flooding of the Wainibuka and Rewa rivers in Fiji in April 2004 damaged between 50 and 70 % of crops (Fiji Government 2004). During the January 2009 floods in Fiji, crops were destroyed and many did not have another source of income to buy food, medical supplies or send children to school (Lal et al. 2009). Drought also presents problems for agriculture everywhere in the region, particularly given the lack of irrigation (Barnett 2007).

River basin modification also affects the food provisioning services that assist communities in the disaster recovery process. Aspects of the relationships between river basin modifications, natural disasters, infectious disease risk and food provisioning are hypothesized in Fig. 3. In a recent Fijian study, a dam constructed in the upper catchment of the Nadi River basin as part of a flood mitigation project (Fig. 4) helped to reduce the frequency of local floods but also resulted in a significant reduction in the availability of fish for local community subsistence (Jenkins and Mailautoka 2011). The dam construction didn't account for the highly migratory nature of insular fish faunas and compounded the long-term nutritional vulnerability of the community.

The highly migratory fish faunas of island systems are disproportionately affected due to a greater probability of encountering obstacles such as dams or other

Fig. 4 A dam constructed to minimize flooding in the upper Nadi Basin in Fiji. Note low water levels and impassability to freshwater fauna. Photo credit: Vinesh Kumar

hydrological modifications, predation by non-native species and degraded water quality. These vulnerable species include ones with the greatest socio-economic value to communities. Jenkins et al. (2010) demonstrated the notable absence from degraded catchments of fishes that traditionally formed staple diets of inland communities. Other notably absent species in heavily modified catchments include many migratory species that form important commercial and cultural fisheries for Pacific islanders. These effects are largely seasonal and magnified in degraded catchments, with pronounced negative impacts during heavy rainfall and severe storms on food-provisioning services and biodiversity (Jenkins and Jupiter 2011). These effects will likely become more severe under predicted future climate scenarios (ABM and CSIRO 2011). Community bans on harvesting and clearing within riparian wetlands can be effective at maintaining fish diversity even in areas where forests have previously been extensively cleared (Jenkins et al. 2010). However, these benefits are rapidly removed once the ban has been lifted and food fish from rivers again become scarce (Jenkins and Jupiter 2011).

Managing river basins to minimize runoff and consequent eutrophication downstream will also provide greater availability of aquatic resources of value such as fisheries. This applies to both within river and downstream coral reef ecosystems. Fabricius (2005) reviewed the negative impacts to coastal coral reefs (e.g. increases in downstream macroalgal cover) associated with impacts of reduced water quality from adjacent, highly modified river basins. Letourneur et al. (1998) showed a decrease in commercial reef fish species richness and biomass with increasing exposure to terrestrial runoff from rivers in New Caledonia. While fishing pressure is certainly a mediating factor in availability of fisheries resources from coral reefs, Wilson et al. (2008) noted from a study in Fiji that habitat loss, in part due to changes in water quality, is currently the overriding agent of change.

Dependence on coral reefs and coral reef fisheries is high in most Pacific SIDS, not only for dietary needs but also for both subsistence and market-based economies (Gillett 2009; Bell et al. 2009). Forty-seven percent of coastal households list fishing as the primary or secondary source of income and in rural communities the subsistence fishery accounts for 60–90% of all fish caught (Gillett 2009). Direct impact on coral reef habitat that supports coral reef fisheries production can have a devastating impact on food security. For example, in Solomon Islands an 8.1 magnitude earthquake followed by a tsunami in April 2007 resulted in rapid, massive uplift of large sections of coral reefs and mangrove ecosystems in areas of Western Province, and severe damage to reefs of western Choiseul from wave energy. Monitoring data from impacted Choiseul coral reef systems show significant decreases in the availability of food fish and invertebrates, as well as a reduction in hard coral cover (Hamilton et al. 2007).

Impacts as Sites of Physical Hazards

At least 1% of the global coastal wetland estate is lost each year primarily by direct reclamation (Nicholls et al. 1999), and these losses may have profound impacts on ecosystem services related to coastal protection. There is a body of evidence suggesting that coastal wetland ecosystems (including mangrove forests, coral reefs and salt marshes) can help to reduce the direct risk of damage and injuries associated with some natural disaster events (e.g. Danielsen et al. 2005; Das and Vincent 2009). The immediate injuries recorded during disasters in the proximity of coastal wetland systems include lacerations, blunt trauma, sprains/strains and puncture wounds, often in the feet and lower extremities, which become susceptible to infection (Ahern 2005; Shultz et al. 2005; Hendrickson et al. 1997). However the precise extent to which mangroves and other coastal vegetation ecosystems serve as bio-shields is debated (Bayas et al. 2011), and the role of coastal wetland vegetation in wave impact and storm surge mitigation still remains controversial (Geist et al. 2006; Iverson and Prasad 2007; Kaplan et al. 2009).

Data on the extent of coastal vegetation and impacts on Pacific Islands post-tsunami are limited, yet some lessons can be learned from examples from South East Asia. In the coastal regions of western Aceh in 2004, the potential for mitigating tsunami impacts appeared limited as a result of the massive energy released by waves with heights exceeding 20 m (Cochard 2011). Studies do suggest, however, that coastal wetland vegetation appears to reduce casualties and damage (Kathiresan and Rajendran 2005; Bayas et al. 2011). Bayas et al. (2011) reported loss of life decreased by 3–8% behind coastal vegetation, indicating that trees may have slowed and/or diverted the waves, thereby allowing greater opportunity to escape to safety. In contrast, when plantations or forests were situated behind villages there was a 3–5 % increase in the number of casualties (Bayas et al. 2011). Similarly, 2100 people were killed and at least 800 were severely injured following a 1998 tsunami in eastern Saundaun Province, Papua New Guinea, where villages were located directly on sand spits in front of the mangroves surrounding the lagoon (Dengler and Preuss 2003). Given these mixed outcomes, mangrove planting as a tsunami mitigation measure has received criticism for failing to control for confounding factors such as distance from shoreline (Kerr and Baird 2007; Baird et al. 2009). One recent review concluded that the value of coastal vegetation as a tsunami buffer is minor (Baird and Kerr 2008), with the authors suggesting that the claim that coastal wetland vegetation can act as a bio-shield gives false hope to vulnerable communities.

Recent experimental studies show that mangroves can reduce the height of wind and swell waves over relatively short distances (McIvor et al. 2012), though these coastal protection services are likely nonlinear and vary with habitat area and width (Barbier et al. 2008). Wave height can be reduced by 13–66 % over 100 m of mangroves. However, most studies have measured the attenuation of only relatively small waves (wave height < 70 cm) and further research is needed to measure the attenuation of larger wind and swell waves by mangroves. In addition, the ability of mangroves to provide coastal defense services is dependent on their capacity to adapt to projected rates of sea level rise (McIvor et al. 2012). Recent evidence suggests that mangrove surface soils are rising at similar rates to sea level in a number of locations (McIvor et al. 2013). However, data are only available for a small number of sites and mostly over short time periods. To allow for continued mangrove protection, wetland managers need to monitor the conditions allowing mangroves to persist and adapt to changing sea levels. This requires monitoring the maintenance of sediment inputs, protection from degradation and the provision of adequate space for landward migration (McIvor et al. 2012). The removal of coastal wetland vegetation may be a greater short-term threat to other services such as storm surge abatement and fisheries supply than sea level rise. It is important for concerned communities of practitioners to be promoting mangroves and coastal wetland vegetation as more than a bio-shield but also as an important source of livelihood provisioning that can assist in the medium to long-term disaster recovery process (Bayas et al. 2011). Artificial protection structures like seawalls cannot provide the nursery and fisheries benefits of natural systems, such as mangroves and coral reefs that can assist in hastening community recovery.

Impacts on Psychosocial Well-Being

Wetlands are of great cultural and social significance to Pacific Islanders and play a crucial role in mediating psychosocial health and well-being. The use of wetlands and wetland ecosystem services is arguably an important aspect of Pacific cultural identity (Strathern et al. 2002). Pacific authors argue that notions of well-being are linked closely to cultural identity (McMullin 2005), that illness is often seen as an inevitable disruption to life and social systems (Drummond and Va'ai-Wells 2004), and that health gives meaning to an individual's place and actions within a community context (Ewalt and Mokuau 1995). Thus, if Pacific Islanders lose their wetland services, they are at risk of a diminution of cultural identity and social well-being.

Some authors note an ecology driven model of well-being that is based on the vitality and abundance of natural resources relied upon for subsistence and cultural practices (McGregor et al. 2003). Within this ecological model, the collective family unit forms the core social unit within which the individual lives and interacts, which is interdependent upon the lands and associated resources for health (physical, mental and emotional) and social well- being. In both cases, the myriad wetland ecosystems of the Pacific are the settings for health where cultural identity, subsistence life and social systems co-exist (*sensu* Horwitz and Finlayson 2011).

While descriptions of psychosocial issues surrounding disasters in Pacific SIDS are relatively scant, it is well documented that populations exposed to natural disasters are affected by a variety of mental health issues (eg. Cook et al. 2008; Shultz et al. 2005). The exposure to loss of life or loved ones, social displacement and economic loss have wide-ranging health effects and contribute to persistent post-traumatic stress disorder (PTSD) and depression (Cook et al. 2008). Persistent PTSD was documented in New Zealand after Cyclone Bola in 1988 (Eustace et al. 1999). When a succession of five typhoons struck Guam in 1992, persons who had acute stress disorder following the initial typhoon were significantly more likely to have progressed to PTSD or depression after the full series of typhoons (Staab et al. 1996). Increases in the rates of suicide and child abuse have also been reported post-cyclones (Shultz et al. 2005) and significant evidence exists post-flooding that children experience long-term increases in PTSD, depression, and dissatisfaction with life (Ahern et al. 2005). Addressing chronic care injuries and psychosocial well-being in the aftermath of these types of disasters can span several generations. Galea et al. (2005) suggested that over a third of initial cases of post-disaster PTSD can persist for more than a decade. As Pacific SIDS are particularly sensitive to disaster risk given the concentration of people along low-lying, wetland areas (e.g. McIvor et al. 2012), post-disaster mental health issues may also be concentrated in these vulnerable populations.

Well-managed wetlands may help to mitigate the psychosocial stress and mental health issues associated with natural disasters. Damaged wetland systems and the community awareness and perception of this damage may result in pathological mental health outcomes as outlined above. A recent study on Pacific Island families showed that those families located in neighborhoods with perceived higher levels of environmental pollution (including wetland degradation), noise and reduced safety were more than twice as likely to have psychological morbidity, with mothers 7.3 times more likely to be affected (Carter et al. 2009).

While the loss of loved ones, income, livelihood options and displacement all contribute to mental distress, the loss of a "sense of place" is also emerging as an important mental health consideration. Albrecht (2005) has used the particular term "solastalgia" for the pain or sickness caused by the loss or, or inability to derive, solace connected to the present state of one's home environment. This concept of solastalgia is illuminated in any case where the environment has been negatively affected by forces that undermine a personal and community sense of identity, belonging and control (Horwitz et al. 2012). This concept is of particular relevance in the context of natural disasters and for the Pacific Islands where cultural identity and factors of spiritual significance are closely tied with local environmental conditions (McMullin 2005; McGregor et al. 2003).

Evidence is also emerging that people actively involved in local conservation projects report better general health and a sense of community belonging than those who were not involved (Moore et al. 2006). Wetlands managers and public health practitioners need to recognize the need to align wetland conservation and restoration efforts with the potential effects on mental health in both the disaster prevention and recovery contexts. Managing wetlands in the wake of natural disasters to minimize the future damage to livelihoods or exposure to pathogens and improve community solidarity in the recovery process are likely to positively contribute to the mental well-being and the speed of recovery in the affected population.

Managing Wetlands for Disaster Risk Reduction and Public Health

The health consequences of disasters cannot be considered in isolation from the wetland ecosystems in which they occur. Wetlands can help to provide protective and essential provisioning services during and following disasters, though in some conditions, they can also become a setting for disease spread and exacerbate harmful conditions. Evidence-based management of wetland conditions surrounding disaster events is important for both disaster response and prevention well beyond the time and place of the disaster itself. The identification and management of short-term health impacts surrounding disasters often captures the attention and financial assistance while many intermediate to long-term impacts are overlooked (Cook et al. 2008). Recovery from disasters such as tropical cyclones, flooding, or tsunamis around wetland areas is commonly a protracted process. Disaster risk reduction strategies must factor in the medium- to longer-term preventative and recovery processes that integrate traditional public health interventions with wetland conservation and restoration, land use planning and climate change adaptation.

Opportunities and Examples of Good Practice at Regional, National and Local Levels

At a global level, following the World Conference for Disaster Reduction, the United Nations General Assembly produced the Hyogo Framework for Action in 2005, a 10-year framework bringing different sectors and actors under a common global system of coordination to reduce disaster losses (UNISDR 2005). In the context of managing wetlands for public health in the face of disasters, the focus of policy makers seeking to integrate wetland management with disaster risk reduction under the Hyogo Framework should focus on Priority Action 4 *"Reducing the underlying risk factors"* by targeting the appropriate sectoral development planning and programs. Implementing agencies of this framework can act through global alliances such as Partnership for Environment and Disaster Risk Reduction (PEDRR) to advocate the importance of wetland management in achieving health outcomes through Priority Action 4.

At the Pacific regional level, the current policy framework for disaster management is the Pacific Disaster Risk Reduction and Disaster Management Framework (PDRRDMF) for Action 2005–2015. The Secretariat of the Pacific Community's Applied Geoscience Commission works to support countries to adapt this framework and implement priorities at a national and sectoral level. This often comes in the form of developing National Action Plans and more recently Joint National Action Plans which seek to address both disaster and climate change risks (SOPAC 2013). Work is underway to develop a new Strategy for Disaster and Climate Resilient Development in the Pacific (SRDP) to succeed the current PDRRDMF and 'Pacific Islands Framework for Action on Climate Change', both due to expire in 2015. The replacement strategy will combine the two inter-related fields of disaster risk reduction and climate change adaptation. Within the implementation of this framework, there will be opportunities for countries to tailor national strategies around managing and restoring wetlands to reduce disaster risk and improve adaptive capacity of coastal communities.

The Ramsar Convention on Wetlands provides another international policy framework for the management of wetlands and their associated services. Aligned with the Ramsar Convention, the Pacific regional framework for wetlands management is the Regional Wetlands Action Plan for the Pacific 2011–2013. Goal 1.4 calls for *"Precise linkages between ecosystem health and human health in the Pacific to be investigated"* and *"Improved engagement between wetland decision makers and human health sectors"*, but underlying strategies and objectives have not been explicit in this regard. The challenge is for countries and territories to embed specific objectives for wetland management for health outcomes in the implementation of their respective National Biodiversity Strategies and Action Plans, which typically cover commitments under the Ramsar Convention.

Important learning opportunities exist within some local approaches to island wetlands management collated with a handbook of good practice in Pacific Integrated Island Management (IIM) (Jupiter et al. 2013). IIM is an approach that calls for "sustainable and adaptive management of natural resources through coordinated networks of institutions and communities that bridge ecosystems and stakeholders with the common goals of maintaining ecosystem services and securing human health and well-being" (Jupiter et al. 2013). Five of the ten guiding principles of IIM are particularly relevant for those seeking to integrate island wetland management into disaster risk management systems: adopt a long-term integrated approach to ecosystem management; maintain and restore connectivity between complex social and ecological systems; incorporate stakeholders through participatory governance with collective choice arrangements, taking into consideration gender and social equity outcomes; recognize uncertainty and plan for adaptive management through regular monitoring, evaluation and review leading to evidence-based decision-making; and organize management systems in nested layers across sectors, social systems and habitats. Box 1 illustrates these principles applied to the management of the Takitumu Lagoon, Cook Islands, for coupled health and environmental outcomes.

Box 1:   Takimutu Lagoon Health Report Card, Rarotonga

The Takitumu district of the high island of Rarotonga, Cook Islands, has developed an integrated ecosystem-based management plan encompassing high island forests, streams, coastal plains and the coral reef lagoon (Dakers and Evans 2007). Advisory committees were established across several sectors to deliver the components of the management plan including government, donor and local leader steering committees, a technical advisory group for issues surrounding environmental monitoring and an inter-departmental committee for within government coordination. In particular, declining stream and lagoon water quality associated with piggery waste was a focal issue around which environmental and health sector authorities could engage. A Takitumu Lagoon Health Report Card was produced for each village for overall water quality, bacterial load, ciguatera in landed fishes, lagoon faunal abundance, adjacent stream water quality, stream bacterial load and safety of groundwater, and then it was shared widely with communities and relevant stakeholders. As a result of this focused attention on connectivity across wetland systems, new Public Health (Sewage) Regulations and an associated Code of Practice were developed. In addition, improvements were made in the system for assessing and approving changes to existing land use through a tightening of regulations needed for planning consent by the Environment Authority. This example demonstrates how IIM planning across multiple island wetland habitats can successfully bring together a wide range of stakeholders around shared concerns of public health and environmental quality (Jupiter et al. 2013). It also demonstrates that synthesizing high quality technical information around these shared concerns is catalytic in garnering both community support and effecting policy change relevant to both wetlands management and public health, thus strengthening the resilience of the community to environmental exposure to disaster risks.

Conclusions and Further Research Needs

Natural disasters are a continuing and growing threat to public health with particular impact on the small developing nations of the Pacific. Disasters affect human health and the ecological properties of wetlands in ways that are not always obvious or expected in the medium- to long-term. While wetlands can mitigate the effects of some disaster related health issues, they can also exacerbate ill health in affected populations. The extent to which specific wetland contexts and types can facilitate disaster mitigation and recovery is generally poorly studied. Building greater understanding at the interface of wetland ecology and disaster-related epidemiology is needed to strengthen existing models of disaster risk management and wetland conservation. Health surveillance systems should incorporate aspects of wetland quality and key provisioning services alongside routine disease surveillance and continue to collect this information in the weeks, months and years following a disaster. This will provide policy makers and managers the tools to monitor and evaluate the longer acting health consequences of interventions in various wetland disaster settings.

A precautionary approach is encouraged for wetland managers to provide careful evidence-based recommendations with regard to the extent of services to expect from wetlands surrounding natural disaster events. While being cautious with advice, wetland managers and scientists must also be proactive in seeking collaboration with public health and disaster risk management arenas to enhance both prevention and recovery strategies. To help manage the impacts of physical hazards, wetland scientists can determine the extent of damage to wetland systems and assist in adapting built and natural infrastructure around wetlands to minimize risks. Wetland ecologists should assist public health authorities in the identification of environmental reservoirs and exposure routes to infectious diseases for humans, livestock and other wildlife. The traditional core work of protecting and restoring important wetland types, such as riparian vegetation and mangroves, can help mitigate community exposure to water contamination and storm surge and also help provide long term livelihood benefits such as fisheries and psychosocial benefits such as exercise and community engagement.

References

ADB and IFPRI (2009) Building climate resilience in the agricultural sector of Asia and the Pacific. Asian Development Bank, Manila

Adger WN (2006) Vulnerability. Global Environ Change 16:268–281

ADW (2012) World risk report 2012. http://www.ehs.unu.edu/file/get/10487. Accessed 12 Feb 2014

Ahern M, Kovats RS, Wilkinson P, Few R, Matthies F (2005) Global health impacts of floods: epidemiologic evidence. Epidemiol Rev 27:36–46

Albrecht G (2005) Solastalgia: a new concept in human health and identity. Philos Activ Nat 3:41–55

Australian Bureau of Meteorology and CSIRO (2011) Climate change in the Pacific: scientific assessment and new research. Volume 1: regional overview. Pacific Climate Change Science Program, Aspendale

Baird A, Kerr A (2008) Landscape analysis and Tsunami damage in Aceh: comment on Iverson and Prasad (2007). Landsc Ecol 23:3–5

Baird AH, Bhalla RS, Kerr AM, Pelkey NW, Srinivas V (2009) Do mangroves provide an effective barrier to storm surges? Proc Natl Acad Sci U S A 106(40):E111

Barbier EB, Koch EW, Silliman BR, Hacker SD, Wolanski E, Primavera J, Granek EF, Polasky S, Aswani S, Cramer LA, Stoms DM, Kennedy CJ, Bael D, Kappel CV, Perillo GME, Reed DJ (2008) Coastal ecosystem-based management with nonlinear ecological functions and values. Science 319:321–323

Barnett J (2007) Food security and climate change in the South Pacific. Pac Ecol 2007:32–36

Barnett J, Campbell J (2010) Climate change and small island states: power, knowledge and the South Pacific. Earthscan Ltd., London

Bayas JC, Marohn C, Dercon G, Dewi S, Piepho PH, Laxman J, van Noordwijk M, Cadisch G (2011) Influence of coastal vegetation on the 2004 tsunami wave impact in West Aceh. Proc Natl Acad Sci U S A 108(46):18612–18617

Bell JD, Kronen M, Vunisea A, Nash WJ, Keeble G, Demmke A, Pontifex S, Andrefouet S (2009) Planning the use of fish for food security in the Pacific. Mar Policy 33(1):64–76

Below R, Wirtz A, Debarati G (2009) Disaster category classification and peril terminology for operational purposes. WHO Centre for research on the epidemiology of disasters (CRED) and Munich Reinsurance Company (Munich RE) Working paper 264. Université catholique de Louvain, Brussels

Bradshaw CJA, Sodhiw NS, Pehwz KSH, Brook BW (2007) Global evidence that deforestation amplifies flood risk and severity in the developing world. Glob Change Biol 13(11):2379–2395

Cabral JPS (2010) Water microbiology, bacterial pathogens and water. Int J Environ Res Public Health 7:3657–3703

Carter S, Williams M, Paterson J, Iusitini L (2009) Do perceptions of neighbourhood problems contribute to mental health. Findings from the Pacific Islands Families study. Health Place 15(2):622–630

Cherry JA (2012) Ecology of wetland ecosystems: water, substrate, and life. Nat Educ Know 3(10):16

Church JA, White NJ, Hunter JR (2006) Sea-level rise at tropical Pacific and Indian ocean islands. Glob Planet Change 53(3):155–168

Cochard R (2011) On the strengths and drawbacks of Tsunami-buffer forests. Proc Nat Acad Sci U S A 108(46):18571–18572

Cockburn A, Clair J St, Silverstein K (1999) The politics of "Natural" disaster: who made Mitch so bad? Int J Health Serv 29(2):459–462

Colwell RR (1996) Global climate and infectious disease: the cholera paradigm. Science 274(5295):2025–2031

Cook A, Watson J, van Buynder P, Robertson A, Weinstein P (2008) 10th anniversary review: natural disasters and their long term impacts on the health of communities. J Environ Monitor 10(2):167–175

Costa MH, Botta A, Cardille JA (2003) Effects of large-scale changes in land cover on the discharge of the Tocantins River, Southeastern Amazonia. J Hydrol 283(1–4):206–217

Costanza R, D'Arge R, De Groot R, Farber S, Grasso M, Hannon B, Limburg K, Naeem S, O'Neill RV, Paruelo J, Raskin RG, Sutton P, van den Belt M (1997) The value of the world's ecosystem services and natural capital. Nature 387:253–260

Dakers A, Evans J (2007) Wastewater management in Rarotonga: it is not just a matter of a technological fix? In: Patterson RA (ed) Proceedings of the workshop on innovation and technology for on-site systems. Armidale

Danielsen F, Sorensen MK, Olwig MF, Selvam V, Parish F, Burgess ND, Hiraishi T, Karunagaran VM, Rasmussen MS, Hansen LB, Quarto A, Suryadiputra N (2005) The Asian Tsunami: a protective role for coastal vegetation. Science 310(5748):643

Das S, Vincent JR (2009) Mangroves protected villages and reduced death toll during Indian super cyclone. Proc Natl Acad Sci U S A 106(18):7357–7360

Dengler L, Preuss J (2003) Mitigation lessons from the July 17 1998 Papua New Guinea Tsunami. Pure Appl Geophys 160(10):2001–2031

Dennison L, Kiem M (2009) The health consequences of flooding. Stu Br Med J 17:110–111

Dillaha TA, Reneau RB, Mostaghimi S, Lee D (1989) Vegetative filter strips for agricultural non-point source pollution control. T Am Soc Agri Eng 32:513–519

Doocy S, Dick A, Daniels A, Kirsch TD (2013) The human impact of tropical cyclones: a historical review of events 1980–2009 and systematic literature review. PLOS Curr Disaster. doi:10.1371/s2664354a5571512063ed29d25ffbce74

Drummond W, Va'ai-Wells O (2004). Health and human development models across cultures. Nagare Press, Palmerston North

Dupon JF (1986) Atolls and the cyclone hazard: a case study of the Tuamotu islands. South Pacific Regional Environment Programme (SPREP)/South Pacific Commission (SPC), Apia and Noumea

Eisenberg NS, Cevallos W, Ponce K, Levy K, Bates SJ, Scott JC, Hubbard A, Vierira N, Endara P, Espinel M, Trueba G, Riley LW, Trostle J (2006) Environmental change and infectious disease: How new roads affect the transmission of diarrheal pathogens in rural Ecuador. PNAS 103(51):19460–19465

Ellison JC (2009) Wetlands of the Pacific islands region. Wet Ecol Manage 17(3):169–206

Eustace KL, MacDonald C, Long NR (1999) Cyclone Bola: a study of the psychological after-effects. Anxiety Stress Copin 12(3):285–298

Ewalt PL, Mokuau N (1995) Self-determination from a Pacific perspective. Soc Work 40(2):168–175

Fabricius KE (2005) Effects of terrestrial runoff on the ecology of corals and coral reefs: review and synthesis. Mar Pollut Bull 50(2):125–146

Falkland T (1999) Water resources issues of small island developing states. Nat Resour Forum 23(3):245–260

Fiji Government (2004) Preliminary estimates of flood affected areas. Press release 17 April, Suva

Galea S (2007) The long-term health consequences of disasters and mass traumas. Can Med Assoc J 176(9):1293–1294

Galea S, Nandi A, Vlahov D (2005) The epidemiology of post- traumatic stress disorder after disasters. Epidemiol Rev 27(1):78–91

Gaudet JJ (1978) Effects of a tropical swamp on water quality. Verhan Internat Vereinigung Limnol 20:2202–2206

Gaynor K, Katz A, Park S, Nakata M, Clark T, Effler P (2007) Leptospirosis on Oahu: an outbreak associated with flooding of a university campus. Am J Trop Med Hyg 76:882–885

Geist EL, Titov VV, Synolakis CE (2006) Tsunami: wave of change. Sci Am 294:56–63

Gillett R (2009) The contribution of fisheries to the economies of Pacific island countries and territories. Asian Development Bank, Suva

Gubler DJ, Reiter P, Ebi KL, Yap W, Nasci R, Patz JA (2001) Climate variability and change in the United States: potential impacts on vector- and rodent-borne diseases. Environ Health Perspect 109(2):223–233

Gupta A (1988) Large floods as geomorphic events in the humid tropics. In: Baker VR, Kochel RC, Patton PC (eds) Flood geomorphology. Wiley, New York, pp 301–315

Hamilton R, Ramohia P, Hughes A, Siota C, Kere N, Giningele M, Kereseka J, Taniveke F, Tanito N, Atu W, Tanavalu L (2007) Post-Tsunami assessment of Zinoa marine conservation area, South Choiseul, Solomon Islands. TNC Pacific Island Countries Report No. 4/07

Hansen MC, Stehman SV, Potapov PV (2010) Quantification of global gross forest cover loss. Proc Natl Acad Sci U S A 107(19):8650–8655

Harbin M, Faris R, Gad AM, Hafez ON, Ramzy R, Buck AA (1993) The resurgence of lymphatic filariasis in the Nile delta. Bull World Health Organ 71(1):49–54

Harper SL, Friedrich B, Hacker J, Hasnain SE, Mettenleiter TC, Schell B (2012) Climate change and infectious diseases. Epidemiol Infect 140(4):765

Harvell CD, Mitchell CE, Ward JR, Altizer S, Dobson A, Ostfeld RS, Samuel MD (2002) Climate warming and disease risks for terrestrial and marine biota. Science 296(5576):2158–2162

Hendrickson LA, Vogt RL, Goebert D, Pon E (1997) Morbidity on Kauai before and after Hurricane Iniki. Prev Med 26(5):711–716

Horwitz P, Finlayson M (2011) Wetlands as settings for human health: incorporating ecosystem services and health impact assessment into water resource management. Bioscience 61:678–688

Horwitz P, Finlayson M, Weinstein P (2012) Healthy wetlands, healthy people: a review of wetlands and human health interactions. Ramsar Technical Report No 6. Secretariat of the Ramsar Convention on Wetlands, Gland & The World Health Organization, Geneva

IFRC (2013) Emergency appeal: Republic of the Marshall Islands/Pacific drought. Emergency appeal n° MDRMH001, GLIDE n° DR-2013-000053-MHL 20 December 2013

IPCC (2007) Climate change 2007: the physical science basis. Interngovernmental Panel on Climate Change, Geneva, Switzerland

Iverson LR, Prasad AM (2007) Modeling Tsunami damage in Aceh: a reply. Landscape Ecol 23:7–10

Jenkins K (2010) Post cyclone Tomas support to typhoid fever control in Fiji March 2010. Fiji Health Sector Improvement Program, Suva, Fiji

Jenkins AP, Mailautoka K (2011) Fishes of Nadi basin and bay: conservation ecology and habitat mobility. In: Askew N, Prasad, SR (eds) Proceedings of the Second Fiji Conservation Science Forum 2011. Wildlife Conservation Society Fiji Program, Suva

Jenkins AP, Jupiter SD (2011) Spatial and seasonal patterns in freshwater ichthyofaunal communities of a tropical high island in Fiji. Environ Biol Fish 91(3):261–274

Jenkins AP, Jupiter SD, Qauqau I, Atherton J (2010) The importance of ecosystem-based management for conserving migratory pathways on tropical high islands: a case study from Fiji. Aquat Conserv 20(2):224–238

Jha AK, Stanton-Geddes Z (2013) Strong, safe, and resilient: a strategic policy guide for disaster risk management in East Asia and the Pacific. Directions in development. World Bank, Washington, DC. doi:10.1596/978-0-8213-9805-0

Jobin W (1999) Dams and disease: ecological design and health impacts of large dams, canals, and irrigation systems. E & FN Spon, London

Jupiter SD, Jenkins AP, Lee Long WJ, Maxwell SL, Watson JEM, Hodge KB, Govan H, Carruthers TJB (2013) Pacific integrated island management—principles, case studies and lessons learned. Secretariat of the Pacific Regional Environment Programme (SPREP), Apia

Kaplan M, Renaud FG, Lüchters G (2009) Vulnerability assessment and protective effects of coastal vegetation during the 2004 Tsunami in Sri Lanka. Nat Hazard Earth Syst 9:1479–1494

Kathiresan K, Rajendran N (2005) Coastal mangrove forests mitigated Tsunami. Estuar Coast Bhelf Sci 65(3):601–606

Kerr AM, Baird AH (2007) Natural barriers to natural disasters. Bioscience 57(2):102–103

Knutson TR, McBride JL, Chan J, Emanuel K, Holland G, Landsea C, Held I, Kossin JP, Srivastava AK, Sugi M (2010) Tropical cyclones and climate change. Nat Geosci 3(779):157–163

Kouadio IK, Aljunid MS, Kamigaki T, Hammad K, Oshitani H (2012) Infectious diseases following natural disasters: prevention and control measures. Expert Rev Anti-infect Therap 10(1):95–104

Krishnamoorthy K, Jambulingam P, Natarajan R, Shriram AN, Das PK, Sehgal SC (2005) Altered environment and risk of malaria outbreak in South Andaman, Andaman & Nicobar Islands, India affected by Tsunami disaster. Malaria J 4:32

Lal PN, Singh R, Holland P (2009) Relationship between natural disasters and poverty: a Fiji case study. SOPAC miscellaneous report No. 678. International Strategy for Disaster Reduction, Suva

Lata S, Nunn P (2012) Misperceptions of climate-change risk as barriers to climate-change adaptation: a case study from the Rewa Delta, Fiji. Clim Change 110:169–186

Lau CL, Smythe LD, Craig SB, Weinstein P (2010) Climate change, flooding, urbanisation and leptospirosis: fuelling the fire? Trans R Soc Trop Med Hyg 104:631–638

Lau CL, Clements ACA, Skelly C, Dobson AJ, Smythe LD (2012) Leptospirosis in American Samoa—estimating and mapping risk using environmental data. PLoS Negl Trop Dis 6(5):e1669. doi:10.1371/journal.pntd.0001669

Lechat MF (1976) Disaster epidemiology. Ann Soc Belge Med Trop 56(4–5):193–197

Letourneur Y, Kulbicki, Labrosse P (1998) Spatial structure of commercial reef fishes along a terrestrial runoff gradient in the northern lagoon of New Caledonia. Environ Biol Fish 51:141–159

Likens GE, Bormann FH, Johnson NM, Fisher DW, Pierce RS (1970) Effects of forest cutting and herbicide treatment on nutrient budgets in the Hubbard Brook watershed-ecosystem. Ecol Monogr 40(1):23–47

Lipp EK, Huq A, Colwell RR (2002) Effects of global climate on infectious disease: the cholera model. Clin Microbiol Rev 15(4):757–770

Lowrance RR, Altier LS, Newbold JD, Schnabel RR, Groffman PM, Denver JM (1997) Water quality functions of riparian forest buffer systems in the Chesapeake Bay Watershed. Environ Manage 21:687–712

Malek EA (1975) Effect of Aswan high dam on prevalence of schistosomiasis in Egypt. Trop Geogr Med 27(4):359–364

Maltby E, Acreman MC (2011) Ecosystem services of wetlands: pathfinder for a new paradigm. Hydrolog Sci J 56(8):1–19

Mackenzie JS, Gubler DJ, Petersen LR (2004) Emerging flaviviruses: the spread and resurgence of Japanese encephalitis, West Nile and dengue viruses. Nat Med 10(12):98–109

Mackenzie E, Prasad B, Kaloumaira A (2005) Economic impact of natural disasters on development in the Pacific. University of the South Pacific, Suva

McGregor DP, Morelli PT, Matsuoka JK, Rodenhurst R, Kong N, Spencer MS (2003) An ecological model of native Hawaiian well-being. Pac Health Dialog 10(2):106–128

McIvor AL, Möller I, Spencer T, Spalding M (2012) Reduction of wind and swell waves by mangroves. Natural Coastal Protection Series: Report 1. Cambridge Coastal Research Unit Working Paper 40. The Nature Conservancy and Wetlands International

McIvor AL, Spencer T, Möller I, Spalding M (2013) The response of mangrove soil surface elevation to sea level rise. Natural Coastal Protection Series: report 3. Cambridge Coastal Research Unit Working Paper 42.The Nature Conservancy and Wetlands International

McKergow LA, Weaver DM, Prosser IP, Grayson RB, Reed AEG (2003) Before and after riparian management: sediment and nutrient exports from a small agricultural catchment, Western Australia. J Hydrol 270(1):253–272

McMullin J (2005) The call to life: revitalizing a healthy Hawaiian identity. Soc Sci Med 61(4):809–820

Millennium Ecosystem Assessment (2005) Ecosystems and human well-being: synthesis. Island Press, Washington, DC

Moore M, Townsend M, Oldroyd J (2006) Linking human and ecosystem health: the benefits of community involvement in conservation groups. Ecohealth J 3(4):255–261

Mosley LM, Sharp DS, Singh S (2004) Effects of a tropical cyclone on the water quality of a remote Pacific island. Disasters 28(4):393–405

Myers SS, Patz J (2009) Emerging threats to human health from global environmental change. Ann Rev Env Res 34:223–252

Myers SS, Gaffikin L, Golden CD, Ostfeld RS, Redford KH, Ricketts TH, Turner WR, Osofsky SA (2013) Human health impacts of ecosystem alteration. Proc Natl Acad Sci U S A. www.pnas.org/cgi/doi/10.1073/pnas.1218656110

Nicholls RJ, Hoozemans MJF, Marchand M (1999) Increasing flood risk and wetland losses due to global sea-level rise: regional and global analyses. Glob Env Change 9(1):69–87

Nicholls RJ, Wong PP, Burkett VR, Codignotto J, Hay JE, McLean RF, Ragoonaden S, Woodroffe CD (2007) Coastal systems and low-lying areas. In: Parry ML, Canziani OF, Palutikof JP, van der Linden PJ, Hanson CE (eds) Climate change 2007: impacts, adaptation and vulnerability. Cambridge University Press, Cambridge

NIWA (2012) Island climate update 140. http://www.niwa.co.nz/climate/icu/island-climate-update-140-may-2012. Accessed 12 Jan 2014

Norris V (1993) The use of buffer zones to protect water quality: a review. Water Resour Manage 7:257–272

Ostfeld RS (2009) Climate change and the distribution and intensity of infectious diseases. Ecology 90(4):903–905

Patz JA (2000) Climate change and health: new research challenges. Ecosyst Health 6(1):52–58

Patz JA, Epstein PR, Burke TA, Balbus JM (1996) Global climate change and emerging infectious diseases. J Am Med Ass 275(3):217–223

Patz JA, Daszak P, Tabor GM, Aguirre AA, Pearl M, Epstein J (2004) Unhealthy landscapes: policy recommendations on land use change and infectious disease emergence. Environ Health Perspect 112(10):1092–1098

Pimentel D, Allen J, Beers A, Guinand L, Hawkins A, Linder R, McLaughlin P, Meer B, Musonda D, Perdue D, Poisson S, Salazar R, Siebert S, Stoner K (1993) Soil erosion and agricultural productivity. In: Pimentel D (ed) World soil erosion and conservation. Cambridge University Press, Cambridge

Ragosta G, Evensen C, Atwill ER, Walker M, Ticktin T, Asquith A, Tate KW (2010) Causal connections between water quality and land use in a rural tropical island watershed. Ecohealth 7(1):105–113

Roth CH (2004) A framework relating soil surface condition to infiltration and sediment and nutrient mobilization in grazed rangelands of northeastern Queensland, Australia. Earth Surf Proc Land 29:1093–1104

Salinger MJ, Lefale P (2005) The occurrence and predictability of extreme events over the Southwest Pacific with particular reference to ENSO. In: Sivakumar MVK, Motha RP, Das HP (eds) Natural disasters and extreme events in agriculture. Springer, Berlin, pp 39–49

Scobie H (2011) Preliminary report: impact assessment of the 2010 Mass Typhoid Vaccination Campaign, Republic of Fiji. Centers for Disease Control and Prevention, World Health Organization, Fiji Ministry of Health, Australian Agency for International Development, Suva, Fiji

Sherif MM, Singh VP (1996). Saltwater intrusion. In: Singh VP (ed) Hydrology of disasters, vol 18. Klewer Academic Publishers,, Dordrecht, pp 269–319

Shultz JM, Russell J, Espinel Z (2005) Epidemiology of tropical cyclones: the dynamics of disaster, disease, and development. Epidemiol Rev 27(1):21–35

Singh RBK, Hales S, de Wet N, Raj R, Hearnden M, Weinstein P (2001) The influence of climate variation and change on diarrheal disease in the Pacific Islands. Environ Health Persp 109:155–159

SOPAC (2013) Regional progress report on the implementation of the Hyogo Framework for Action (2011–2013). Secretariat of the Pacific Community, Suva

Staab JP, Grieger TA, Fullerton CS (1996) Acute stress disorder, subsequent posttraumatic stress disorder and depression after a series of typhoons. Anxiety 2(5):219–225

Strathern A, Stewart PJ, Carucci LM, Poyer L, Feinberg R, Macpherson C (2002) Oceania: an introduction to the cultures and identities of Pacific Islanders. Carolina Academic Press, Durham

Terry JP, Falkland AC (2009) Responses of atoll freshwater lenses to storm-surge overwash in the Northern Cook Islands. Hydrogeol J 18(3):749–759

UNDP South Pacific Office (2002) Natural disaster risk reduction in Pacific Island countries, final report for international decade for natural disaster reduction 1990–2000. United Nations Development Programme, Fiji

United Nations International Strategy for Disaster Reduction (UNISDR) (2005) Hyogo Framework for Action 2005–2015: building the resilience of nations and communities to disasters. http://www.unisdr.org/wcdr. Accessed 10 Jan 2014

Walzer PD, Judson FN, Murphy KB, Healy GR, English DK, Schultz MG (1973) Balantidiasis outbreak in Truk. Am J Trop Med Hyg 22(1):33–41

Watson JT, Gayer M, Connolly MA (2007) Epidemics after natural disasters. Emerg Infect Dis 13(1):1–5

Wilson SK, Fisher R, Pratchett MS, Graham NAJ, Dulvey NK, Turner RA, Cakacaka A, Polunin NVC, Rushton SP (2008) Exploitation and habitat degradation as agents of change within coral reef fish communities. Glob Change Biol 14(12):2796–2809

WHO (2013) Outbreak surveillance and response priorities for mitigating the health impact of disaster. Tenth Pacific health ministers meeting. Agenda Item 9. http://www.wpro.who.int/southpacific/pic_meeting/2013/documents/PHMM_PIC10_9_Outbreak. Accessed 10 Jan 2014

Young S, Balluz L, Malilay J (2004) Natural and technologic hazardous material releases during and after natural disasters: a review. Sci Total Environ 322(1–3):3–20

Interventions Required to Enhance Wetlands as Settings for Human Well-Being

Pierre Horwitz, C Max Finlayson and Ritesh Kumar

Abstract The close relationship that exists between food production, water use and water extraction dictates that those charged with wetland management consider a broader societal objective that extends beyond nature conservation and a sectoral or simplistic approach to natural resource management. Given the interactions that occur between wetlands and human health is it imperative that wetland managers are involved in efforts to build and sustain the coping capacity of affected human communities, and to recognize that these efforts will need to operate at local, national, or regional levels. This is because the factors, such as poverty and high burdens of disease, that place populations at risk can also limit the capacity of these populations to prepare for the future, or in this instance, make wise use of their wetland ecosystems. In proposing ways in which societies can intervene via wetland management we draw significantly on material presented in a similarly titled chapter of a Ramsar Technical Report, and from a Resolution passed by the Ramsar Convention on Wetlands 2012 Conference of Parties in Romania. Many of the possible response options for addressing ecosystem change and human health and well-being lie outside the direct control of the wetland and water sectors, or indeed, even the health sector. Instead they are likely to be embedded in areas such as sanitation

P. Horwitz (✉)
School of Natural Sciences, Edith Cowan University, Joondalup, WA, Australia
e-mail: p.horwitz@ecu.edu.au

C. M. Finlayson
Institute for Land, Water and Society, Charles Sturt University, Albury, Australia
UNESCO-IHE, Institute for Water Education, Delft, The Netherlands
e-mail: mfinlayson@csu.edu.au

R. Kumar
Wetlands International—South Asia, A 25, Second Floor, Defence Colony, 110024 New Delhi, India
e-mail: ritesh.kumar@wi-sa.org

© Springer Science+Business Media Dordrecht 2015   193
C. M. Finlayson et al. (eds.), *Wetlands and Human Health,* Wetlands: Ecology, Conservation and Management 5, DOI 10.1007/978-94-017-9609-5_10

and water supply, education and training, agriculture and fisheries, trade, tourism, transport, development, housing and infrastructure. As a consequence integrated interventions will need to address existing social values and cultural norms, existing infrastructure, and the social, economic, and demographic driving forces that result in wetland change. This includes ensuring steps are taken to enable marginalised stakeholders to be effectively represented at all stages of the management cycle, increased transparency and access to information, and engaging with and supporting the core pursuits of other sectors. Possible responses could encompass steps to: promote cross-sectoral governance and institutional structures; promote and rationalize incentive structures; support social and behavioural responses, including capacity building, communication and empowerment; develop technological solutions such as a way of enhancing multi functionality of ecosystems; and develop other cognitive responses.

Keywords Poverty · Burdens of disease · Sectors · Cross-sectoral · Integration · Interventions · Stakeholders · Transparency · Governance · Incentives · Behaviours · Capacity building · Communication · Empowerment · Technological solutions · Health costs · Deliberative democracy · Disciplines · Trade-offs

Introduction

A common theme when discussing human well-being has been an emphasis on the strong interdependence between ecosystems, including wetlands, and human health as an important part of human well-being (MEA 2005, Horwitz et al 2012, Patz et al. 2012). Horwitz et al. (2012) presented an analysis of this interdependence for wetlands by considering: (a) the linkages that occur between the ecological character of a wetland and its ecosystems services, (b) the way in which ecosystem services benefited human well-being, (c) the drivers of ecosystem change that diminished the contributions of those ecosystem services, and (d) the outcomes and effect of such changes on human health. Arising from this analysis was an imperative to consider what interventions were required to enhance human well-being by addressing the erosion of ecosystem services in wetlands.

As wetland and human health are both being affected by the same spectrum of direct and indirect drivers of change, many of which are global in nature, there is a pressing need to address them to help achieve mutual outcomes—improved health outcomes for people and the maintenance or enhancement of ecosystem services from wetlands (Horwitz and Finlayson 2011, Finlayson and Horwitz 2015). The global issues needing attention are introduced in, for example, the Millennium Ecosystem Assessment (MEA 2005) and the Global Environment Outlook (UNE 2012). It is clear that increasingly urgent changes in societal attitudes and governmental policies and responses are required to address these within wetlands, including at a local scale. There is a range of available or promising response options that may help. These issues are further examined below, drawing heavily on an initial examination in a technical report published by the Ramsar Convention on Wetlands

(Horwitz et al. 2012) and recognizing wetlands as settings for human well-being (Horwitz and Finlayson 2011).

There are several critical points here that have been used to determine the structure of this chapter. For reasons presented in the Global Environment Outlook (Armenteras and Finlayson 2012) and The Economics of Ecosystems and Biodiversity (Russi et al. 2013), it is clear that a new approach is needed to address the magnitude of cross-sectoral challenges. Hence, in line with the context presented in Horwitz et al. (2012) we deal with *attitudinal shifts and reorientation of perspectives to enable those with a wetland and human health question to construct their problem statements*. In addressing these issues we draw significantly on a similarly titled chapter in Horwitz et al. (2012), and from a Resolution adopted by the Ramsar Convention's 2012 Conference of Parties in Romania (Ramsar Convention 2012).

The new approach being considered as a way to address the mutual outcomes for wetlands and human health and well-being will require a reformulation of the water and wetland management agenda in many countries along with adjustments within governments to develop cross-sectoral approaches to societal matters, be they ones of ecosystem management or restoration, supporting public health, or changes to agriculture or public infrastructure. Without new policy initiatives and further consideration of the consequences for governance, it will be extremely difficult to address on-site wetland management and public health issues in a coordinated manner. The second section of this chapter will seek to *explain how particular interventions at a higher level of policy development will enable on-ground action*.

In response to the above imperatives wetland managers will be encouraged to consider how their actions will also affect human health. Conversely, ecosystem disruption could occur if human health and well-being matters were considered in isolation. Wetland managers will therefore need to consider the positive or negative effect of their actions on wetland ecosystems and on human health and well-being. The last section of the chapter will *provide some instruments and approaches that will allow wetland managers to assess the possible implications of their actions on human health and well-being*.

The interventions that are presented cover policy and practice and were derived from recent international investigations, in particular the Millennium Ecosystem Assessment (MEA 2005), the Global Environment Outlook (UNEP 2007, 2012), the 2nd UN World Water Development Report (2006) and the Comprehensive Assessment of Water Management in Agriculture (Molden 2007). Further, the Health in All Policy (Kickbusch et al. 2008; Adelaide Statement on Health in all Policies 2010) was also used to demonstrate the locus of some of these issues within the wider health sector.

Changing Attitudes and Perspectives

(Horwitz et al. 2012) point out that the above mentioned compendia and policy agendas repeatedly emphasise two important points about ecosystem change and human well-being: *proposed response and intervention themes are often common across different sectors*, yet many of the possible *response options to human health and ecosystem change lie primarily outside the direct control of the wetland sector, and the health sector.* Rather they are embedded within other sectors, including sanitation and water supply, education, agriculture, trade, tourism, transport, development, and housing. Because of this they placed a lot of emphasis on the importance of identifying key partners and stakeholder groups needed to achieve appropriate outcomes. To be most effective, wetland policy-makers will need to 'create a space' to enable integration across partners and other interest groups to ensure the potential health impacts of ecosystem change are reduced. Integrated approaches, and the creation of spaces for partners and stakeholders are part of the mix needed to address existing social values and cultural norms, existing infrastructure and the social, economic, and demographic driving forces that result in ecosystem change. These driving forces don't just produce changes in ecosystems, they are themselves a product of them in a reciprocal and interdependent manner.

Using systems thinking, wetland managers realize that there are consequences of their actions, and undertake these actions *knowing* about them, notwithstanding the fact that they live in a complex and uncertain world. (Horwitz et al. 2012) suggested the following four attitudinal changes to assist wetland managers realize the consequences of their actions:

I.   Where trade-offs are being made, they need to be considered and valued according to principles of sustainability and equity rather than ignored or dealt with based on partial economic or financial terms only.

II. It is not acceptable to reason that we can manage wetlands for biodiversity alone; in fact to do so, as argued in this chapter, will be counterproductive. A people-centred approach in wetland management, based on a conceptualization of these ecosystems as coupled social ecological systems can help in identifying pathways which do not diminish the importance of biodiversity, also help achieving co-benefits of sustainable ecosystem management and the Millennium Development Goals (MDGs). ["Social ecological systems" make explicit the complex linkages between human behaviour and organisation, and the biophysical world. In fact the linkages are so intertwined that the system becomes 'self organised' to deal with the phenomena that emerge from the relationships.]

III. Resolving matters of trade-offs across levels of human involvement from the personal to the global, is achievable with dialogue, using a deliberative rather than hierarchical approach, to ensure that more powerful forces do not marginalize the local interests of people.

IV. Identifying principal partners and responsible stakeholder groups, often across disciplines and between sectors where barriers and boundaries exist, requires a particular form of engagement that wetland managers need to develop as part of their skill set: patience, tolerance of these 'others' and a willingness to reciprocate.

The attitudes and perspectives outlined in the text above are shown pictorially in Fig. 1.

Fig. 1 Attitudes and perspective shifts required for wetland managers to ensure that the human health consequences of their actions are adequately considered in policy and practice. (Adapted from Horwitz et al. 2012)

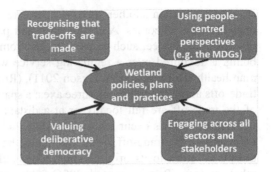

Thinking About Trade-Offs

The Comprehensive Assessment of Water Management in Agriculture (Molden 2007) identified a number of *big* trade-offs, all controversial and potentially provocative, and inherently paradoxical. One of them comprises a trade-off between human populations whereby the need to address and solve particular human health issues could lead to an increase in population and place further pressure on local resources, potentially causing other forms of human suffering. Trade-offs are a common feature of the complexity that characterizes social ecological systems and can result in either virtuous or vicious cycles with the former comprising societal benefits that reinforce the decision-making, and the latter comprising decision-making that makes societal conditions worse. Resolving the intricacies associated with human population trade-offs cannot be done through simple formulae, instead flexible, innovative and adaptive approaches are needed.

A further trade-off identified by the Comprehensive Assessment (Molden 2007) can occur when equity and productivity issues are considered together. In particular the trade-off occurs when the promotion of equitable, nutritious and environmentally benign agriculture is not as productive as more efficient or production-oriented forms of agriculture that tend to favour the wealthy or larger agricultural enterprises rather than the poorer members of society. This situation can also lead to an intergenerational trade-off whereby tensions occur between providing for the quality of life for the current generation at the expense of the next, or indeed, vice versa.

High on the list of trade-offs identified by Molden (2007) was the need to determine whether water should be stored for agriculture or be reallocated for environmental purposes, largely to help restore rivers and their floodplains. A further part of this trade-off comprises decisions about how much water should be allocated to either, and whether or not a reallocation would result in an overallocation and have effects upstream or downstream. Hence, trade-offs about water allocations also encompass decisions about the reallocation or overallocation of water and to what extent decisions are needed about upstream causes and downstream effects, and who benefits or loses.

A key trade-off when considering the interactions between people and wetlands is where ecosystem services that support particular health or well-being outcomes are disrupted when another set of services are enhanced to favour different health or well-being outcomes. A conspicuous and paradoxical example occurs when a provisioning service, such as using water from a wetland to benefit human well-being, is favoured over a regulating service with negative consequences for human health (Horwitz and Finlayson 2011). (Rodriguez et al. 2006) classified such trade-offs along the following three axes: a spatial axis to show whether the effects of the trade-off are felt locally or at a distant location; a temporal axis to show whether the effects occur relatively rapidly or slowly; and an axis to show the reversibility of the trade-off, or the likelihood that a service will be extinguished and unable to return to its original condition should management activities focus on other services. (Rodriquez et al. 2006) further emphasized the importance of politicians, regulators, and the public understanding the consequences of trade-offs and of taking one path in preference to another. Recognising the potential for trade-offs is the important first step in understanding and modeling consequences under different scenarios, for each of the axes outlined by (Rodriguez et al. 2006). Once the potential for trade-offs has been recognized the central concern is the *process* by which the trade-offs and their consequences are negotiated, including ensuring the inclusion of components to support transparency and access to information, as well as engagement with the core interests of other relevant sectors, and enabling marginalised stakeholders to be represented.

Contributions to the Millennium Development Goals

The complementarities that exist between ecosystem services and the benefits they provide for human health and well-being are achievable when wetland managers and policy-makers make a contribution towards the Millennium Development Goals (MDGs). This is particularly the case when considering the close relationships that occur between food production, hunger and poverty, climate change, water use and extraction, and wetland management, as shown in Table 1 (which was adopted from Horwitz et al. 2012) and compiled using material presented in (Molden 2007; UNEP 2007; UN WWDR 2006) and as otherwise indicated).

The Millennium Development Goals (MDGs) were adopted by the United Nations in 2000 through the *Millennium Declaration* with the aim of improving the lives of people around the world, particularly the most vulnerable and disadvantaged. The MDGs included specific targets for the year 2015 to help lift people out of poverty, save lives, ensure adequate childhood education, reduce maternal deaths, and expand opportunities for women and girls through empowerment. As pointed out by Horwitz et al. (2012) the MDGs are directly relevant to the management of wetlands and water resources with a focus on ensuring access to clean water and alleviating the burden of deadly and debilitating diseases. They further promote sustainable development and protect the most vulnerable from the devastating effects of multiple crises (United Nations 2011); targets that are intricately associated with the wise use concept of the Ramsar Convention (Finlayson et al. 2011)

Table 1 Ways in which wetland management can contribute to the achievement of the Millennium Development Goals. (from Horwitz et al. 2012) with minor modifications)

Millennium Development Goals (MDGs)	How will intervening in disruption to wetland ecosystem services improve human health and help address the MDGs?	Systemic consequences: where will addressing MDGs need to be aware of the relationship between human health & wetland health?
1. Eradicate extreme poverty & hunger	Food security of the poor often depends on healthy ecosystems & the diversity of ecosystem services they provide. Diverse wetland ecosystems are self-sustaining & provide the essential genetic material for aquaculture & horticulture. Sustainable livelihoods by definition seek to ensure that the core requirements of food & water are provided to those dependent on the provisioning of wetland ecosystems	The challenge for irrigated agriculture is to improve equity, reduce environmental damage, increase ecosystem services, & enhance water & land productivity in existing & new irrigated systems (Molden 2007). Improving productivity should not be at the expense of other ecosystem services. If it is, the human, animal & plant health consequences of ecosystem disruption will occur in full or in part due to a range of both direct & indirect impacts, the latter as a result of altered health status of livestock & wildlife
2. Achieve universal primary education	Wetland management needs to address the disruptions to ecosystem services that result in water-related diseases Water-related diseases such as diarrheal infections cost about 443 million school days each year, diminish learning potential & reduce the coping capacity of local populations for current predicaments & future ecosystem changes	Primary education should include knowledge of health, water & energy issues at least (a fundamental necessity for urban dwellers who have become more alienated from their surroundings than at any stage in history). Education services can have tendencies to resist increases in attention to such environmental issues at the expense of other subjects
3. Promote gender equality & empower women	Addressing degradation in wetlands, such as water contamination & deforestation, will contribute to the health of women & girls. Women & girls bear the brunt of collecting water & fuelwood & are more vulnerable members of populations to water-borne diseases	Improved wetland management should involve women & girls in a meaningful way, perhaps by recognizing that women can play greater roles in wetland management than they currently do. "Wetland managers", as professions, tend to be dominated by men. Decision-making structures for water resource management, wetland management, & agriculture are also gendered in many parts of the world. These may operate as barriers to achieve this Goal

Table 1 (continued)

Millennium Development Goals (MDGs)	How will intervening in disruption to wetland ecosystem services improve human health and help address the MDGs?	Systemic consequences: where will addressing MDGs need to be aware of the relationship between human health & wetland health?
4. Reduce child mortality	Wetland management will become an essential operational requirement to reduce exposures to waterborne diseases, such as diarrhoea & cholera. Prevalence of these diseases is a result of disruption of regulatory services (as a result of over-extraction & inappropriate practices)	Interventions such as water treatment facilities (often through aid provision) will usually be technological & infrastructural in the short term to address immediate needs. However the medium- to long-term goal should be the management of wetland ecosystems to ensure that they can provide suitable water purification & pathogen removal services
5. Improve maternal health	Addressing disruptions to wetland ecosystem services will always include an examination of water quality. Provision of clean water reduces the incidence of diseases that undermine maternal health & contribute to maternal morbidity & mortality	Improving the quality of source water from catchments, reservoirs & wetlands in general, & distribution infrastructure, may reduce disinfection loads & the likelihood of maternal exposures to these loads
6. Combat major diseases	Up to 20% of the burden of disease in developing countries may be associated with environmental risk factors. Preventive environmental health measures are as important & at times more cost-effective than health treatment. Managing wetlands to enhance ecosystem services with the aim of reducing the likelihood of human exposures to pollutants & infectious diseases is preventive, attending to upstream environmental determinants of health. New biodiversity-derived medicines hold promises for fighting major diseases	Increasing human population sizes as a consequence of successful disease prevention measures may also increase pressure on local water & wetland resources. Wetland management needs to act in concert with water resource management to deal with these foreseeable consequences, for instance by increasing awareness & thus changing behaviour, & by incorporating the concept of ecosystem services in prevention strategies. This management needs to be integrated with regional population policies, domestic livestock & wildlife policies (to reduce risk of emerging zoonoses), education & awareness

Table 1 (continued)

Millennium Development Goals (MDGs)	How will intervening in disruption to wetland ecosystem services improve human health and help address the MDGs?	Systemic consequences: where will addressing MDGs need to be aware of the relationship between human health & wetland health?
7. Ensure environmental sustainability	Current trends in environmental degradation need to be reversed in order to sustain the health & productivity of the world's ecosystems. Wetlands, & the biodiversity they support, encompass many of the key ecosystems of the world & many of the most productive ones. Wetland management applies directly to this Goal	Development strategies that aim to safeguard the full range of benefits provided by wetlands might better achieve the Goal while minimizing harm to wetlands. This requires recognizing the trade-offs that exist when managing for some ecosystem services like those concerned with production, while trading-off supporting & regulating services
8. Develop a global partnership for development	Poor countries need to exploit their natural resources, like wetland ecosystems, to raise revenue to meet debt repayments. Inequitable globalization practices result in harmful side effects in countries that often do not have sufficiently effective governance regimes to counter these	Trade, tourism & migrations of species are often transcontinental. Meaningful wetland management acknowledges that pests & pathogens capable of decreasing ecosystem services & having consequences for the health of local human, domestic & wildlife communities can be distributed by inappropriately planned & controlled human activities. This needs appropriate recognition in global partnerships for development

including when addressing climate change (Finlayson 2011) and poverty reduction (Kumar et al. 2011).

Attempts to achieve the goals and targets of the MDGs have received more attention as shown by Horwitz et al. (2012) with the following quotation "At the 2010 High-level Plenary Meeting of the General Assembly on the Millennium Development Goals, world leaders reaffirmed their commitment to the MDGs and called for intensified collective action and the expansion of successful approaches" (United Nations 2011, p. 5]. Contributions by the wetland sector to achieving the MDGs were foreseen as occurring along two general directions (Horwotz et al. 2012). The first entailed interventions to halt the ongoing disruption to wetland ecosystem services so as to provide opportunities to help to improve the health of humans, and that of domestic and wild species (see column two in Table 1). The second comprised an analysis of the systemic consequences of steps for achieving the MDGs (see column three in Table 1). In particular this recognized that interventions in support of the MDGs, improving human health, and enhancing the ecosystem services from wetlands may not be mutually beneficial—instead systemic effects, such as cross-scale interactions and the consequences of feedbacks may undermine or even negate the original objectives. Thus activities by the international and national communities and by sectors other than wetland management need to be aware of and understand the systemic nature of the relationships. They also need to be aware of the paradoxical outcomes that can exist between these objectives and maintaining the health of wetland ecosystem (Horwitz and Finlayson 2011). The potentially negative consequences of these paradoxes are often foreseeable, and are not necessarily a reason to avoid actions to achieve the MDGs; rather, they provide a signal to understand the consequences and to consider them in decision making.

An understanding of the consequences of the trade-offs that occur among different wetland ecosystem services and the need for collaboration and cooperation across sectors is critical for determining and designing actions that can support the MDGs. As an example, and as documented in the Millennium Ecosystem Assessment (2005) it is not uncommon for activities that are designed to increase food production and reduce poverty to also result in the conversion of marshes to agriculture and mangroves to aquaculture, as well as lead to a significant increase in the use of agrochemicals to increase production. At the same time these activities will lead to a reduction in habitat quality and area and the magnitude of the ecosystem services provided by the original habitat, as well as an increase in water pollution, the loss of the water filtering service formerly provided by the wetlands, and the removal or diminishment of other ecosystem services provided by mangroves, including protection from storm surges, the supply of charcoal and timber for construction, and the degradation or loss of fish habitat, on which local residents may have relied for everyday purposes. These outcomes, even if inadvertent could impede steps to achieve the goal of improved water and sanitation, and potentially increase poverty by undermining some of the benefits formerly available to some groups. In contrast, an approach that safeguarded the range of benefits provided by the wetlands may have more beneficial outcomes and minimise harm to the wetlands.

Deliberation in Managing Trade-Offs

As many ecosystem services are seen as public goods that provide benefits to society there is widespread acceptance that society overall is better served if they are maintained, even if there are only small numbers of people who individually or collectively, or privately and exclusively benefit from them (MEA 2005). This raises concerns, both ethically and normatively, about the social rights and wrongs concerning their management, including the role and importance of dialogue in fully appreciating the impacts of alternate management choices. These issues were raised by Horwitz et al. (2012) who recognized that the heterogeneity of power structures, unequal social positions and differential strengths in political bargaining processes could inhibit rational decision making and lead to marginalization and the creation of imposed choices. They further supported the general principle that wherever common goals were needed they should be worked out in a manner and by processes wherein each individual was fairly represented with deliberation playing an important role in managing trade-offs. As deliberation requires individuals to move beyond private self-interest there is a greater likelihood of achieving outcomes based on both social equity and political legitimacy (Elster 1997). Unfortunately value-laden trade-offs and cost-benefit analysis often fall short of delivering social equity.

In an effort to increase deliberation and ensure participation in decision making, a variety of approaches have been designed and employed, ranging from consultation as an information gathering exercise to full engagement with responsibility for making decisions (Arnstein 1969). Stakeholder engagement is one of the most commonly used techniques, often with small focus groups being used to obtain deliberative solutions. Other authors have recommended using a mix of approaches, such as social multi-criteria analysis and mediated modelling (Kallis et al. 2006, stagl 2007).

Though appealing, deliberation, especially stakeholder engagement, is not always simple. Stakeholders can be difficult to engage (Artz 2005), especially where they have power within existing processes and want to protect this, or have prior expectations from the institutions involved. Further to these difficulties wetland managers and health service providers need to accept that the outcomes from different consultation processes may not be the same, and have a strong sense of realism about the social nature of local politics in particular. Strong willed and clever opinion makers can readily lead dialogue in specific directions, and consensus can be a tool to silence dissent or limit the input of specific groups. In some instances it may be as useful to explain, explore and respect rather than remove dissident views in an attempt to achieve consensus (Spash 2007).

Engaging with Other Sectors

A sector is generally seen as a level of society that shares a common set of goals, represented by its own language, and with agreed approaches and behaviours. In this respect, the *wetland sector* will be different to the *health sector*, and the *private sector* will be different to the *public sector*. Further, the different levels of government are represented by different sectors—the local and the national government sectors. Each sector tends to develop their set of responsibilities in a hierarchical manner with resources distributed and power relationships established accordingly. They further develop their own languages and patterns of behaviour around their core business, although they may overlap, at times to a considerable extent. As an example, water, sanitation and hygiene are often referred to as a sector with overlaps with other sectors, including public works, energy and nutrition. In order to manage the global disease burden associated with the water, sanitation and hygiene sector the other sectors must be engaged to act at both a policy level and on their specific activities. These other sectors manage or have influence over the management of both the determinants of health (e.g. managing the water levels in a dam) as well as their direct actions (e.g. maintaining safe water and sanitation in workplaces), but differ in other spheres of their activity and emphasis. Another example is provided by the overlapping and different responsibilities of the water resources and the wetland sectors.

Governance structures are the main mechanisms for maintaining these sectors and imposing boundaries that need to be negotiated between sectors, with the boundaries being most expressed at 'middle-management' where sectoral territories are most defended. Closer to the operational level where interaction with local interests is likely to be easier there are often on-the-ground and action-oriented practices that already operate across sectors; these should be encouraged and supported through governance structures and policies. There may though be a problem of circularity where there can be a reinforcing and almost irreconcilable tension between maintaining the sanctity of the territory (the sector) and the practicality of getting the job done.

Cross-sectoral actions need to overcome the boundaries that are created (and often defended) between sectors, necessitating reciprocity, mutual understanding and respect. The imperative for wetland policy-makers is to establish ways in which they can engage with the health sector, while understanding the demands that society places on the health sector, and articulating wetland management needs using an inclusive, perhaps even co-constructed, language, with the assumption that the engagement will be reciprocated. Identifying common areas and overlap provides an opportunity to construct a potentially powerful way of establishing links between sectors as disparate as wetlands and public health. An agenda that supports the co-benefits of healthy ecosystems and healthy people was identified through the settings approach (Horwitz and Finlayson 2011) and is elaborated below.

Policy Level Responses and Interventions

Horwitz et al. (2012) provided a synthesis of some proposed policy interventions of relevance for enhancing wetland ecosystem services and human health based on information derived from previous and recent international investigations. In most cases the policy interventions that were identified were non-specific and non-targeted. Their application can be broad provided that the context is taken into account and the particularities are specified. The responses that were identified ranged from: promoting cross-sectoral governance and institutional structures; promoting rationalized incentive structures; social and behavioural responses which included capacity building, communication and empowerment; and technological solutions for enhancing the multi-functionality of ecosystems.

The information from the synthesis was used to identify appropriate mechanisms that would enable health costs to be satisfactorily included into wetland management (Fig. 2).

Fig. 2 Policy shifts and interventions to enable wetland practices to accommodate the notions associated with ecosystem services and human health. (Adapted from Horwitz et al. (2012))

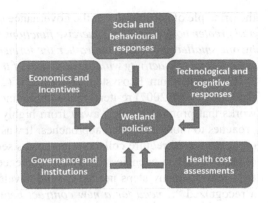

Institutions and Governance

As discussed above (Horwitz et al. 2012) have identified the considerable potential that exists to build on currently used social frameworks that focus on deliberative and collective, often involving multiple numbers of stakeholders, approaches to support the objectives of achieving both public health and ecosystem management objectives. There is a consensus that to meet these objectives society needs to develop governance processes based on adaptive management, social learning, and cross-sectoral engagement in institutional agendas. This is consistent with the view expressed in the Adelaide Statement on Health in all Policies (2010) that boundaries between sectors, disciplines, communities and cultures can be transcended through place-based settings for social learning and action.

Society will also need to develop suitable institutional and governance frameworks that can build trust, social cohesion and reduce inequalities if watershed management is to both enhance ecosystem services and improve the environmental and social determinants of health. While the design principles for long-enduring natural resource management organisations identified by Ostrom (1990) has relevance for developing such frameworks Falkenmark and Folke (2002) have summarised a set of governance imperatives for this purpose:

- securing of social acceptance of measures that are considered necessary and limit the earlier degrees of freedom;
- arrangements for resolution of dispute between stakeholders with incompatible interests;
- attention to existing nestedness between both catchments and subcatchments on the one hand and between ecosystems on the other.

These imperatives are reflected in the World Water Development Report (2006) that deals with steps needed to promote and protect human health; in particular it recommended that the starting point for governance needed for the planning, development and management of water resources at the river basin level was the multiple uses and multiple users of water. It also recommended the promotion and adoption of the principle of subsidiarity in the governance of water resources whereby *a larger and greater body should not exercise functions which can be carried out efficiently by one smaller and lesser, but rather the former should support the latter and help to coordinate its activity with the activities of the whole community*; (Mele 2004).

The Millennium Ecosystem Assessment (2005) and the Global Environment Outlook (UNEP 2007) re-iterated the importance of developing institutional frameworks that promoted a shift away from highly sectoral resource management approaches to more integrated approaches. It was also recognized that integrated approaches would require collaboration across sectors within government, and coordination between agencies to ensure coherence in international negotiations, and then into the many steps needed for the development of national policy. There is a recognized "...*need for a new contract between all sectors to advance human development, sustainability and equity, as well as improve health outcomes. This requires a new form of governance where there is joined-up leadership within governments, across all sectors and between levels of government*." (Adelaide Statement on Health in all Policies (2010); see also Kickbusch 2008).

An interesting theme identified from the abovementioned reports was the parallel recognition by the environment sector and the health sector of the need to develop ways to respectively integrate ecosystem management goals and public health goals within other sectors and within broader development frameworks. Similarly, the environment and health sectors were both making broadly-based calls in support of greater transparency and accountability from government and private-sector organisations that were involved in making decisions that affected ecosystems and human health, including an emphasis on the increased involvement of stakeholders in decision-making.

Integrated and adaptive management approaches were seen as fundamental in achieving social and economic development goals, rather than single issue, command-and-control regulatory approaches, when working for the sustainability of aquatic ecosystems to meet the water resource needs of future generations. To be effective, such approaches needed to consider the linkages and interactions that occurred between hydrological systems that extended across multiple "boundaries," whether geographic, political or administrative. Ecosystem-or nature-based management approaches also provide a basis for cooperation in addressing common water resources management issues, and further provide a way to prevent such issues from becoming sources of conflict between countries or regions (Schoeman et al. 2014).

Economics and Incentives

Assessments of human behaviour and the development of incentive systems have been central to societal efforts to ensure the provision of wetland ecosystem services. Valuation is an important means of expressing linkages of human societies with wetlands, their ecosystem services and biodiversity (see Box 1). Given the values people hold for wetlands efforts to make the link between the economy and the environment more explicit could have important outcomes for ecosystem services.

Box 1: On Valuation

The term 'value' has been ascribed a range of meanings, with a particular emphasis on economic valuation in the context of considering wetlands and health interactions. In particular the emphasis is on a subset of valuation approaches which reflects the choices people make about financial and natural resources given a multitude of socio-economic conditions as preferences, distribution of income and wealth, the state of natural environment, production technologies and expectations of the future (Barbier et al. 2009).

Economics uses valuation approaches, such as those described by de Groot et al. (2006) for wetlands, to provide society with information on the relative scarcity of resources. Economic valuation, which is an anthropocentric and utilitarian subset of valuation approaches, is used to help unravel the complexities of socio-ecological relationships, make explicit how decision making affects the value of ecosystem services, and to express changes in these values in units that enable their incorporation into public decision making (Mooney et al. 2005). However, recognizing and being aware of economic values is only the first step towards attitudinal changes for wetlands. Given the public good characteristics of most of the ecosystem service categories, mechanisms and management approaches are required for capturing economic values in real-life decision making.

Market-Based Instruments (MBIs) can play an important role in decision making processes by integrating the costs associated with the loss of value from degraded ecosystem services and using these to influence the behaviour of citizens and companies. There is an increasing level of experience that has shown that positive incentives for continual innovation and improvement can be created by the use of well-designed market based instruments to support environmental goals at less cost than many conventional "command and control" approaches (Stavins 2000). An increased emphasis on market-based or financial instruments within the conservation and development sectors is a part of an expanding shift in policy based on the success that markets can have in inducing changes in behaviour at an individual and institutional level in a cost effective manner.

Horwitz et al. (2012) have pointed to the growing focus on the use of the economic contributions from ecosystem services to design incentive systems wherein external beneficiaries can make direct, contractual and conditional payments to local landholders for adopting practices that ensure continued provision of these services (see for example, Wunder (2005). These mechanisms are increasingly being grouped under the general term of 'Payments for Ecosystem Services' (PES) and represent one of the important developments for linking the valuation of ecosystem services to incentive systems that promote the sustainable use of ecosystem services. Their application to date has been largely focused on carbon sequestration, watershed management and biodiversity conservation (see Box 2 for examples). Through an emphasis on stakeholder-led management and sustainable financing they are increasingly providing a significant tool for addressing wetland loss and degradation. Many of the services that have been considered for PES schemes occur at a landscape scale, with many provided by watersheds and wetlands.

The payments from such schemes, as with the development of water markets and water-pricing, will need to occur in conjunction with steps to improve or change the allocation of rights to freshwater resources to align the payments with conservation needs (Finlayson et al. 2005), and the elimination of subsidies that promote excessive use of ecosystem services, and, where possible, enable the subsidies to be transferred to support those services that are not readily included within market mechanisms (MEA 2005).

The Millennium Ecosystem Assessment (2005) identified a number of other economic instruments and market mechanisms with the potential to enhance the management of ecosystem services including the imposition of user fees or taxes on activities with "external" costs or trade-offs that are not accounted for in the market, the creation of markets, including those dependent on a cap-and-trade systems, and mechanisms that encouraged consumer preferences to be expressed through markets.

Poor communities in developing countries face particular challenges for investing in wetland ecosystem services if they are entangled in a poverty trap: in order to meet short term and critical livelihood needs they can be forced to unsustainably exploit their environment, eroding the critical life support services such as fisheries, timber, soil fertility and freshwater provision, and entrenching the living conditions associated with their poverty. In response to these dire circumstances a biorights financing mechanism that links poverty alleviation and environmental conservation has been developed (van Eijk and Kumar 2009). In return for the provision of microcredits, local communities are encouraged to engage in ecosystem protection and restoration of the processes that support ecosystem services, with the microcredits being converted into definitive payments on successful delivery of services. Successful applications of biorights include mangrove restoration in Java, tsunami affected areas in Aceh, and waterbird conservation in Inner Niger Delta, Mali.

While such approaches are promising and have been increasingly promoted, some recent developments in institutional economics have challenged the perception that markets provide an optimal mechanism for resource allocation, and instead consider markets as just one mechanism in amongst a multitude of institutional and cooperative arrangements and hierarchies that can be used to guide decision making and resource allocation (North 1990; Williamson 1985; Stiglitz 1986). Thus, market based instruments are just part of the range of options that are available to communities, decision makers and policy planners. This provides opportunities for engaging with the economic sectors that can influence risk distribution within society. As an example, the role of wetlands as natural infrastructure that can buffer at least some of the impacts of uncertainty imposed by climate change and anthropogenic pressures alike could be included in risk mitigation strategies; given that the benefits and risks were better communicated and known to key players. Similarly, financial arrangements could be used to ensure the transfer of a segment of the benefits derived by economic sectors from the functioning of healthy ecosystems, into their long term conservation and management, and possibly restoration.

Valuation thereby constitutes an important component of strategies to address wetlands—human health interlinkages, including by acting as a tool for self-reflection, and alerting us to the consequences of our choices and behaviour on the various dimensions of human capital (Zavestoski 2004). In this sense valuation is an important institutional mechanism to engender change in the way societies respond to the problems caused by the ongoing loss and degradation of wetlands. However, in order to play this role efficiently, the valuation processes need to be credible tools for achieving economic efficiency, but also for ensuring social fairness and ecological sustainability. Various authors have commented that the outcomes of any valuation process depend on what stakeholders value, whose values count and why, who benefits, and the manner in which the interlinkages between social and ecological are accounted. It is also increasingly recognized that the values and the processes used for valuation reflect socially and culturally constructed realities that are linked to worldviews, mind-sets and belief systems shaped by social interactions, as well as political and power relations operating within a realm of local, regional and global interdependencies (Wilk and Cliggett 2006; Hornborg et al. 2007). It is thereby pertinent that ecological as well as social contexts are adequately considered and captured in approaches for assessing and capturing values.

Box 2: Examples of Public Private Partnerships for Water Quality—Payments for Ecosystem Services (from Horwitz et al. (2012))

"In 1992 the Société des Eaux Minérales d'Evian joined with the French government to found an organisation to protect the catchment area of the natural spring experimenting with more environmentally friendly farming practices, expanding the nearby sewers, and ensuring regulatory compliance by livestock holdings. The Société now supports two thirds of the conservation costs in the catchment. Similarly Vittel pays US$ 230/ha to farmers to support sustainable agricultural practices within the catchments of its water sources; this is cheaper than constructing filtration units (Smith et al. 2006; Perrot-Maître 2006). The Los Negros scheme in Bolivia focuses on watershed and biodiversity protection where the Pampagranade Municipality pays Santa Rosa farmers for forest and páramo conservation (Asquith et al. 2008). The central government of China initiated the Sloping Land Conversion Program focussing on watershed protection where the central government pays rural households for cropland retirement and afforestation (Bennett 2008)."

Social and Behavioural Approaches

Wetland managers, particularly those with prior experience, will know that different approaches are available when planning or implementing an intervention, generally involving different instruments and forms of engagement. Choosing the most appropriate approach may in some instances be at least as important as the expected outcome of the intervention. For instance plans to improve water sanitation are more likely to succeed following the appropriate participation of parents, particularly women, in local communities in the planning and implementation of these plans. Most, if not all, interventions of this nature will involve steps that affect the water resource itself.

Falkenmark and Folke (2002) emphasised the importance of social learning in water-related matters, including the roles of participation, empowerment, communication and education. They further directed attention away from seeing water-related matters as technical issues in isolation of other social factors. Ivey et al. (2004) proposed five questions that could help outline the capacity of a particular community to deal with water-related matters (specifically climate-induced water shortages):

- Are community stakeholders aware of the potential impacts of water shortages on human and ecological systems?
- Are local water management agencies perceived by community stakeholders as legitimate?
- Do local water management agencies and related organizations communicate, share information, and coordinate their activities?
- Is there an agency providing leadership to local water management organizations?
- Are members of the public involved in water management decision-making and implementation of activities?

The factors that place populations at risk (such as poverty and high burdens of disease) will in many cases also impair the capacity of these populations to respond in an adequate manner for the immediate and longer time periods. Wetland managers therefore need to be involved in building the coping capacity of such populations and recognise that appropriate responses will occur across community, national or regional levels.

There is an increasing realization that the limitations of traditional regulatory approaches have led to the introduction of approaches based on more participation, such as demand management and voluntary agreements. To be successful these will necessitate education and public involvement, including calls for greater literacy about water management: the latter will require public education curricula at all levels to vigorously address the issues of the water environment, and be supported by awareness campaigns (Falkenmark et al. 2007). In line with deliberative approaches mentioned above more attention should be directed towards the manner in which stakeholders are involved. An educated and more involved population will be more effective in addressing the weaknesses or even failures of government and better able to hold institutions to account. Similarly, the empowerment of groups particularly dependent on ecosystem services or affected by their degradation, including women, indigenous people, and young people, will improve the likelihood of better management of the ecosystems that provide those services (MEA 2005).

Technological and Other Cognitive Responses

As part of the Comprehensive Assessment of Water Management in Agriculture Falkenmark et al. (2007) argued that outdated and outmoded irrigation systems needed to be reconfigured and adapted in order to support steps to reduce poverty in rural areas, to keep up with global demand for agricultural products, including adapting to changing food preferences and societal demands, to adapt to urbanization, industrialization, and increasing allocations to the environment, and to respond to climate change. Technologies that increased crop yields without harmful impacts related to water, nutrient, and pesticide use (MEA 2005) were seen as necessary for the future sustainability of agriculture.

Establishing programmes to restore ecosystem services in these ways was a major recommendation from the Millennium Ecosystem Assessment (2005). In making this recommendation it was recognised, for example, that ecosystem restoration could reduce the incidence of some water-borne diseases, but could also lead to an increase in the incidence of others (Horwitz and Finlayson 2011). This negative aspect could be countered by improved understanding of the ecological requirements of disease vectors, and by incorporating this knowledge into wetland restoration projects, and adapting technical approaches accordingly. The 2nd UN World Water Development Report (2006) recommended the following three interventions at the policy level, namely the need to: (i) introduce the use of available tools for estimating costs and benefits of different drinking water and sanitation options initially at the national and subsequently at lower levels of governance; (ii) promote intervention studies that provide scientific information and help strengthen the evidence base on the effectiveness of environmental management methods for control of water-associated vector-borne diseases, and develop a toolkit for environmental managers in this area; and (iii) refine the correlations between water indicators and the indicators for childhood illness/mortality and nutritional status, the importance for accelerated access to safe water and adequate sanitation, and better Integrated Water Resources Management (IWRM) practices.

Evaluating Wetland Management Interventions on the Basis of Health and Well-Being Outcomes

Horwitz et al. (2012) proposed that a particular and critical approach to managing wetlands for both their ecosystem services and for human health required the development of evaluation processes for interventions that appropriately accounted for both. Hence, the evaluation of wetland management interventions needed methods that could be used to value the specific health and well-being outcomes obtained by maintaining or restoring ecosystem services. While theoretical frameworks have been developed for assessing the costs of health outcomes due to the degradation and disruption of ecosystem services these have not been widely applied to wetlands. Conventional practice separates health effects into two broad categories—mortality and morbidity. There is a growing realization of the need and benefits that could come from applying these frameworks to other health outcomes.

Mortality

Economic approaches to valuing reduced mortality are based on the trade-offs made by individuals or government policy makers in relation to the probability of death and other goods having monetary value. The willingness of an individual to pay for a reduction in the probability of death, or the willingness to accept compensation where the probability may be increased has been proposed as a basis for valuation (Schelling 1968; Bailey1980). An individual's willingness to pay (WTP) for changes in the probability of his or her death can be translated into a more convenient figure for evaluating policies that reduce or increase the risk of death by estimating a value of statistical life or the value of statistical deaths avoided. The willingness to pay approach focuses on a reduction in the probability of death being avoided.

The human capital approach provides an alternate approach to willingness to pay. It considers the loss of output and productivity as a consequence of the shortening of life of an individual. It diverges from approaches that consider the individual centric basis of decision making to societal well-being by using income as an indicator of capital, which in essence reflects how society in general perceives the importance of the contribution of an individual.

The Monetary Value of Reduced Morbidity

The monetary value of reduced morbidity could similarly be derived through the use of approaches based on individual preferences (willingness to pay or required compensation), or by using resource or opportunity costs. Damage function relationships can be derived from estimates of the real cost of illness in the form of lost or reduced productivity and output and an increase in the cost of resources needed for medical care. Despite being relative simple to calculate and apply this approach does not include an evaluation of pain and suffering, which could be readily captured through willingness to pay approaches. For valuation purposes, the acute effects are usually modelled and estimated as though they are certain to be avoided, whereas the chronic effects are treated using a probabilistic approach used for mortality.

Apart from monetary valuation, other approaches have been proposed for assessing the health outcomes of policies, including the dollar value for a Quality Adjusted Life Year (QALY). QALYs are generally converted to dollars using a single US$ per QALY factor. The resultant monetary estimates of the benefits can then be used in a cost benefit analysis. Alternately, an estimate can be derived from a set of conversion factors that are used for any particular composition of health effects that is embedded in QALYs.

Impacts of Wetland Policies on Human Health

A review of the literature reveals a gap when it comes to assessment of the health outcomes associated with wetlands. In contrast, many studies have been directed towards evaluating the impacts of policies related to the maintenance of air quality in developed countries (i.e. Krupnic (2004)) and to food-borne illness (see Mauskopf and French 1991). Assessments of the value of a statistical life for the US have ranged from US$ 5–6 millions in different policy contexts with some attempts having also been made to extend the approach to developing countries (Viscusi 1998; Miller 2000). Gyrd-Hansen (2003) estimated the willingness to pay per QALY of DKK 88,000 on the basis of elicited preferences of health status.

There have been attempts to analyse individual behaviour in response to an undesirable health condition as basis for assessing the economic value of health related outcomes. (Harrington et al. 1989) measured losses caused by an outbreak of waterborne giardiasis in Luzerne County, Pennsylvania and concluded that individual households spent US$ 485–1540 to avoid contaminated water. Legget and Booksteal (2000) showed that a change in the concentration of fecal coliforms (of 100 colony forming units of water per 100 mL) affected the sale prices of properties by 1.5 %, with the dollar amounts ranging from US$ 5000–10,000. (Boyle et al. 1999) estimated the demand from households for clear water in lakes and observed a loss of value of at least US$ 25,000 per household following a decline in Secchi disc clarity from 3.78 to 2.41 m. However, these studies do not make a direct linkage between the ecosystem services and the health conditions—making this an obvious and important area for future research.

Despite the reputed robustness of theoretical frameworks for assessing health outcomes there are a number of issues that underpin successful application of valuation and assessment approaches. When asserting that an individual has a WTP for a reduction in the probability of his or her mortality and/or morbidity there is an accompanying assumption that the individual can perceive changes in her/his health status. It has been shown, however, that individual behaviour differs significantly when dealing with voluntary and involuntary risk (Starr 1976). Similarly, there have been debates about the philosophical foundations of the concept of the value of statistical life. For example, it is contended that if life itself is priceless, the risk of a change in status would tend to infinity. Similarly, there have been debates over the probability range which is used for the estimation of the value of statistical life. Typically, it is expected that one would be dealing with lower probabilities of death in most environmental cases, whereas most of the studies tend to use values at the higher end. There are issues related to inter-age variation, latency, and choice of discount rates that pose significant challenges for the method.

Success Factors and Stumbling Blocks in Policy Reforms

In order for the above responses to be effected, some strategies for their implementation are needed, including analyses of what has made particular policies either successful or not. Citing a range of sources the Global Environment Outlook (UNEP 2007) summarised the wide range of factors that have been demonstrated to be important in determining which polices have been better:

- solid research or science underpinning the policy;
- high level of political will, usually bipartisan and therefore sustained;
- multistakeholder involvement, often through formal or informal partnerships;
- willingness to engage in dialogue with policy opponents;
- robust systems for mediating conflict;
- capable, trained staff engaged in implementation;
- prior systems of monitoring and policy revision agreed, including clauses that mandate periodic revision;
- legislative backing, combined with an active environmental judiciary;
- sustainable financing systems, ring-fenced from corruption;
- evaluation and assessment of policies independent from the rule-making agent, for example, by advisory committees or public auditors;
- minimal delays between policy decisions and implementation; and
- coherence and lack of conflict throughout all government policies.

Conversely, (Molden 2007) discussed a number of reasons why past policy initiatives had fallen short, principal among them were:

- policy reforms have not taken into account the history, culture, environment, and vested interests that shape the scope for institutional change;
- policy reforms have been based (only) on "blueprint" solutions—solutions that follow a model that may have been successful elsewhere;
- there is a focus on a single type of organization rather than the larger institutional context;
- reforms have ignored the many other factors that affect water use in agriculture—policies and government agencies in other sectors, informal user institutions, and the macroeconomic environment and broader social institutions;
- any or all of the following are in operation:
  - Inadequate support for reform at required levels.
  - Inadequate capacity building and incentives for change.
  - Repeated underestimation of the time, effort, and investment required to change.

Wetland-Based Interventions—The Role of the Wetland Managers

(Corvalan et al. 2005) pointed out that intervening to reverse the impacts of ecosystem disruptions, even when well-intentioned, may not necessarily have a positive effect on human health. They extracted from the Millennium Ecosystem Assessment a number of responses that had been recommended for dealing with ecosystem disruption and demonstrated that in almost every category of ecosystem response the consequences for health could be either positive or negative. As also demonstrated by (Horwitz and Finlayson 2011) when considering the paradoxical situations that can arise, the reciprocal can also be true—that responses specifically designed to address human health may have either positive or negative consequences for the maintenance of ecosystem services. Corvalan et al. (2005) further suggested that the outcomes from such responses would depend on how the policy or regulation was framed and whether local circumstances and contingencies were taken into account.

From the above discussion (Horwitz et al. 2012) raised some important questions for wetland managers. Firstly, what are the human health consequences of intervening for wetland management? Secondly, what are the local particularities of each wetland ecosystem that may have consequences for human health? Thirdly, how might wetland managers intervene to improve human health itself? In response they argued that wetland managers needed to include these questions in the full suite of considerations that accompanied their decision making about such issues.

In particular, wetland managers needed to be familiar with and use some of the key approaches, tools and instruments that were likely to be used by the health sector to respond to the health effects and health outcomes caused by disruptions to wetland ecosystem services. Commonly used approaches, tools and instruments include monitoring, surveillance and intervention, burden-of-disease assessments (BDA), health impact assessments (HIA), community health assessments, risk assessments, community and stakeholder engagement (see Box 3). In a general sense these might all have environmental equivalents, so conceptually they may be familiar to trained environmental managers; their health focus however, could be very different, and may require adaptation for specific wetland applications.

In particular, packages for HIA and BDA designed for those who may not necessarily have training in health-related sciences could be very useful, especially in situations where management interventions are proposed for water resources or wetlands. Given the diverse nature of impact assessment instruments that are used across social, economic, environmental and health domains, the co-benefits that can be derived from the development of instruments that can be used across these domains, without a loss of detail and analytical emphasis, should be obvious. (Harris-Roxas and Harris 2012) developed a typology of health impact assessments and recognize four specific types:

I. **Mandated HIA** which is done to meet a regulatory or statutory requirement;
II. **Decision support HIA**, which is done voluntarily with the goal of improving decision-making and implementation;
III. **Advocacy HIA** which is conducted by organizations or groups who are neither proponents or decisionmakers, with the goal of influencing decision-making and implementation, and
IV. **Community-led HIA** where potentially affected communities examine issues or proposals that they onsider of concern for their health consequences.

Seen from these perspectives, (Horwitz et al. 2012) proposed a model, in the form of a three-legged stool, for risk management, with each 'leg' being critical to successful wetland management: 1. mitigation of hazard; 2. regulation of behaviour; and 3. education for awareness-raising about the consequences of behaviour and responses.

Each 'leg' makes an essential contribution to addressing the vulnerability of the population at risk and their adaptation options. Each 'leg' further comprises a class of response options or human actions, including policies, strategies, and interventions, to address specific issues, needs, opportunities, or problems.

In the context of ecosystem management, response options may involve governance, institutional, legal, technical, economic and financial changes, or changes in behavior and/or attitudes related to knowledge and awareness. Based on the response options interventions are needed and should be designed at spatial and temporal scales appropriate to the ecosystem disruption and the health outcome of concern; they can address local, national, regional, and international scales and within any of these scales, address specific vulnerable subgroups. Factors that affect the choice of responses include the knowledge and understanding of the underlying processes or causes; the capacity to predict, forecast, and warn; the capacity to respond (institutional and otherwise); how the risk might change over time and with ecosystem change; and ethical appropriateness (Corvalan et al. 2005).

As argued by Cook and Speldewinde (Cook and Speldewinde 2015) the health sector will normally try to establish the evidentiary basis for the spread of a particular diseases, the risk factors for that disease, and ideally, evaluate the outcome from the intervention. This is usually done using the following pathway: monitoring and surveillance of the disease and risk factors; interpretation of data; use of the data in conjunction with environmental and other data to develop models to predict disease occurrence; link changes in disease rates to specific environmental factors; and intervene to remove the causes of disease or to lessen the damage they cause (see Corvalan et al. 2005). Interventions can be evaluated using a similar process.

Box 3: Examples of Public Private Partnerships for Water Quality—Payments for Ecosystem Services (from Horwitz et al. 2012)

In order for any research on the health effects of ecological change to affect either official policy or individual behaviour, it is necessary to take into account how risk is perceived. A deliberate and well-informed approach to community risk will maximize the chance of effective changes through policy interventions that enjoy popular support. Any assessment of ecological change and health should be influenced by the risk perceptions of those communities that are most likely to be affected. That is, ecological assessments should involve open and frequent stakeholder participation from the beginning of the process rather than as an afterthought. This approach of community engagement in the process serves the purpose of accessing local knowledge about the effects of ecological factors, ensuring that the assessment addresses issues of greatest concern to those affected and maximizing the probability that any recommended change in policy or behaviour will be adopted. If a source of information is not widely trusted, it is unlikely that recommended changes will be accepted. Community surveys have shown that some groups tend to be regarded as highly trustworthy, while others (such as government agencies) are treated with caution. Healthcare providers (for instance community nurses or doctors) tend to be one of the "high trust" groups, underlining again the important role they can play in explaining the significance of healthy ecosystems. Any such consultation should make the best use of the expertise of both stakeholders and researchers. Stakeholders may have expert local knowledge but may have inaccurate ideas of the true nature of risks associated with different factors; researchers should have more exact knowledge of disease processes and relative risks but may inappropriately estimate the applicability of general concepts to local situations. Accurate and accessible reporting of assessment results can remedy inaccurate risk perceptions and can enhance the public's ability to evaluate science/policy issues; the individual's ability to make rational personal choices is enhanced. Stakeholder engagement will make it more likely that the research is credible and is translated into practice. Technically intensive, externally driven interventions may produce rapid results but risks marginalizing local communities. Interventions that engage local communities and transfer expertise are more likely to result in longer term ecologically and socially sustainable improvements.

One strategy that may valuably support efforts to ensure 'cross-over' between the wetland and health sectors is the use of data on human disease burden as an indicator to help focus and prioritise wetland restoration. The idea is not new and may support more initiatives, such as using mosquito-borne disease data as a bio-indicator for ecosystem health (Jardine et al. 2008). As data on human health are generally collected more widely and consistently than are data on wetland ecosystems there could be more mutual benefits from closer collaborations between relevant wetland ecologists and health researchers. Such collaboration could help efforts to support the sustainable provision of ecosystem services from wetlands. In some circumstances, particularly when community livelihoods and well-being and wetlands are interdependent and interconnected, data derived from health indicators could be used to reflect the status of wetlands. This provides an opportunity to include health indicators within the suite of indicators that are commonly used to assess the effectiveness of wetland management, particularly where it has direct implications for human health. The specific challenge in this proposal will be to incontrovertibly link the indicator (human disease) with the exact cause of adverse change in the wetland ecosystem both temporally and spatially; there is little evidence that this idea has been evaluated.

Another valuable strategy is to consider whether wetland management plans can encompass the health of local communities, especially in instances where liveli-hoods are clearly dependent on the wetland. This is particularly important given the increasing acceptance that poor health can severely undermine the capacity of communities to maintain systems that support the sustainable management of natural resources and the wise use of wetlands. It is therefore likely that effective community-led wetland management can be ensured when the people who manage them are themselves healthy. For example, in several societies, the role of women in managing wetlands is very important, including, for example, the collection of water, fish and aquatic plants etc., and gives them a particular role in ensuring the health of the wetlands that provide these resources. Wetland managers could there-fore consider steps to ensure that health related services are integrated within the wetland communities as part of the wider intervention strategies adopted for wet-land restoration. Better targeting and participation in the implementation of man-agement plans could be achieved through a more specific focus on the importance of addressing gender and marginalized communities, and thereby contribute to the MDGs, as mentioned above.

In many cases health authorities and experts will know exactly what needs to be done to improve human health, or to prevent a further decline, but the role of wet-land managers in support of such efforts might not be exactly clear. In response to this situation (Horwitz et al. 2012) added an action or a role that wetland managers could adopt to help deliver the response imperatives identified by (WEHAB 2002a, b) for *Water and sanitation* to improve human health. (Box 4).

Box 4: Response Imperatives for Water and Sanitation to Improve Human Health (from). From (Horwitz et al. 2012) Based on Information Derived from (WEHAB 2002a, b). An Action or a Role for Wetland Managers has been Added to each Response

1. Assign the role of water-related public awareness to the agency responsi-ble for integrated water resource management at the country level; *Action: wetland managers need to develop in-house capacities to deliver water-related public awareness.*

2. Institute gender-sensitive systems and policies; *Action: wetland managers examine their internal operations to ensure gender-sensitivity; this should include equity in decision making capacities, particularly where inequali-ties exist in health outcomes.*

3. Raise awareness and understanding of the linkages among water, sanitation, and hygiene and poverty alleviation and sustainable development; *Action: wetland managers develop their own conceptual models for how these linkages can be articulated in national, regional and local contexts.*

4. Develop in partnership with all relevant actors community-level advocacy and training programs that contribute to improved household hygiene practices for the poor; *Action: wetland managers participate in such partnerships when approached.*

5. Identify best practices and lessons learned based on existing projects and programs related to provision of safe water and sanitation services focused on children; *Action: wetland managers identify documentation that demonstrably links management of wetlands to improving ecosystem services and provision of safe water and sanitation relevant for local context.*

6. Create multistakeholder partnership opportunities and alliances at all levels that directly focus on the reduction of child mortality through diseases associated with unsafe water, inadequate sanitation, and poor hygiene; *Action: wetland managers participate as knowledge providers for diminished ecosystem services that result in proliferation of disease conditions.*

7. Develop national, regional, and global programs related to the provision of safe water and improved sanitation services for urban slums in general, and to meet the needs of children in particular; *Action: wetland managers contribute local examples of links between management of wetlands to improving ecosystem services and provision of safe water and sanitation, through their national, regional and global networks*

8. Identify water pollution prevention strategies adapted to local needs to reduce health hazards related to maternal and child mortality; *Action: wetland managers develop particular communication materials and provide advice on the water quality aspects that require preventative strategies.*

As a consequence of the steps outline above it is possible that policy level responses used for one circumstance or purpose can be converted to actions or practices that are appropriate for wetland managers in general. While many interventions will have general application across other sites or regions, the complexity of responses required for individual diseases must not be underestimated. In many case these will need to be developed on a case-by-case basis, as exemplified by the management of avian influenza (HPAI H5N1) (see Box 5), and applied at a local or provincial level, generally with national or international support.

Box 5: Avian Influenza and Wetlands: Appropriate Responses (prepared by Rebecca Lee, David Stroud & Ruth Cromie and Included in (Horwitz et al. 2012)

"As well as providing conditions for virus mutation and generation, agricultural practices, particularly those used within wetlands, can enhance the ability of a virus to spread. The role of Asian domestic ducks in the epidemiology of H5N1 highly pathogenic avian influenza (HPAI) has been closely researched and found to be central not only to the genesis of the virus (Hulse-Post et al. 2005; Sims et al. 2005), but also to its spread and the maintenance of infection in several Asian countries (Shortridge and Melville 2006). Research in Thailand and Vietnam (Gilbert et al. 2008; Songserm et al. 2006) has demonstrated abundance of free-grazing domestic ducks, human population and rice cropping intensity as significant risk factors for H5N1 HPAI. Gilbert et al. (2006) concluded that in Thailand "wetlands used for double-crop rice production, where free-grazing ducks feed year round in rice paddies, appear to be a critical factor in HPAI persistence and spread". This practice of using flocks of domestic ducks for 'cleaning' rice paddies of waste grain and various pests allows contact with wild ducks using the same wetlands. Additionally, the practice of "farming" wild ducks and geese has provided significant opportunities for transmission of virus between domestic and wild populations.

The impacts of H5N1 on human health, livelihoods, poultry production, and local and national economies have been enormous and concern has remained about a potential pandemic strain mutation. The virus has also killed wild birds in many countries with outbreaks sometimes involving threatened species and/or thousands of mortalities. Wild birds have been involved in the spread of the virus, possibly over long distances however a reservoir of infection of H5N1 circulating in wild populations has never been demonstrated despite global surveillance efforts. Their role in the spread and persistence of infection is dwarfed by that of poultry, with markets and trade providing ready opportunities for regional, national and international spread. Yet wild birds are ungoverned and unaffiliated and thus easy to blame as a source of infection despite a lack of evidence. Responses to H5N1 HPAI in many countries highlighted the risk that loose language about the spread of the disease from those involved in disease control plus ill-informed media reporting, may undo decades of building positive public attitudes towards wetland and waterbird conservation. For example, as HPAI H5N1 spread across central Asia and Europe in winter 2005 and spring 2006, there were reports of wild birds being killed and their nests destroyed, some wetlands were altered, visitor numbers at wetland centres in the UK fell markedly with economic impacts for conservation organisations and changed public attitudes, which encompassed concern and even fear. The double negative impact was clear: not only were wild birds killed by a virus essentially of our own making but wild birds were disproportionately blamed for their part—surely more victim than vector?

The role of the multilateral environmental agreements has been key in promoting appropriate responses to HPAI (Cromie et al. 2011). One of the central obligations of the Ramsar Convention is that Contracting Parties "shall promote the conservation of wetlands and waterfowl by establishing nature reserves on wetlands" and subsequent decisions of the Conference of Parties have stressed the role of these reserves and associated wetland centres in enhancing public awareness of wetlands and communicating the need for waterbird conservation. Resolution IX.23 of the Ramsar Convention on Wetlands states the "destruction or substantive modification of wetland habitats with the objective of reducing contact between domesticated and wild birds does not amount to wise use as urged by Article 3.1 of the Convention". The key to the control of HPAI remains prevention and control in the poultry sector (Greger 2006; GRAIN 2006; Sims 2007) and ornithologists and the conservation community must play their part in this to ensure benefits to all.

Human lives are enriched by birds—contact with—and appreciation of which, is an important element of the well being of those who may otherwise have limited opportunities to interact with wildlife. Getting close to birds brings great pleasure. As the late Janet Kear, life-long waterbird conservationist, once said, "just as you can't sneeze with your eyes open, you can't feed a bird from your hand without smiling." It is crucial that we avoid preventable reactions that might encourage people to stay away from wild birds because of unfounded fears and false perceptions of risk. In the long-term, this could prove greatly damaging to public support for wetland and waterbird conservation.

Currently, wetland health problems are being created or exacerbated by activities such as intensive food production systems and trade, wetland loss or degradation, and introduction of alien species. Ultimately, to reduce risk of avian influenza and other diseases, we need to move to markedly more sustainable systems of agriculture with significantly lower intensity systems of poultry production. These need to be more biosecure, separated from wild waterbirds and their natural wetland habitats resulting in far fewer opportunities for viral cross-infection and thus pathogenetic amplification (Greger 2006). To deliver such an objective in a world with an ever-burgeoning human population, hungrier than ever for animal protein, and with major issues of food-security throughout the developing world, will be a major policy challenge. However, the animal and human health consequences of *not* tackling these issues, in terms of the impact on economies, food security and potential implications of a human influenza pandemic, are quite immense."

Conclusions

Efforts to reduce the potential health impacts of adverse change in wetlands will require the adoption of inter- and cross-sectoral integrated options. It is therefore important that the principal cross-sectoral partners and relevant stakeholder groups required to achieve appropriate outcomes are identified. Integrated interventions will by necessity need to address existing social values and cultural norms, existing infrastructure, and the social, economic, and demographic driving forces that result in wetland change.

Given the interactions that occur between wetlands and human health is it imperative that wetland managers are involved in efforts to build and sustain the coping capacity of affected human communities, and to recognize that these efforts will need to operate at local, national, or regional levels. This is because the factors, such as poverty and high burdens of disease, that place populations at risk can also limit the capacity of these populations to prepare for the future, or in this instance, make wise use of their wetland ecosystems.

As many interventions or responses involve trade-offs it is important that the consequences of taking one path in preference to another are recognized and, where necessary, suitable compensation considered. The important first step is being able to recognize the potential for tradeoffs. Hence, a *process* by which tradeoffs can be negotiated is of central concern with key components including representation of marginalised stakeholders, increased transparency and ready access to information, and engagement with the core pursuits of other sectors.

Managing wetland ecosystem services to improve human health will help achieve the Millennium Development Goals, and the Sustainable Development Goals that are being negotiated to replace them post 2015.

Acknowledgements The authors gratefully acknowledge the opportunity to use content from the Ramsar Technical Report 6, and the Ramsar Convention's Resolution XI.12-e.

References

Adelaide Statement on Health in all Policies (2010) WHO, Government of South Australia, Adelaide

Armenteras D, Finlayson CM (2012) Biodiversity. In: UNEP. Keeping track of our changing environment: from Rio to Rio + 20 (1992–2012). Division of Early Warning and Assessment (DEWA), United Nations Environment Programme (UNEP), Nairobi

Arnstein S (1969) A ladder of citizen participation. J Am Inst Plan 35:216–224

Arzt NH (2005) Development and adoption of a national health information network; response to request for information. HLN Consulting, LLC, San Diego

Asquith N, Vargas MT, Wunder S (2008) Selling two environmental services: in-kind payments for bird habitat and watershed protection in Los Negros, Bolivia. Ecol Econ 65:675–684

Bailey MJ (1980) Reducing risks in to life: measurement of benefits. Report. American Enterprise Institute, Washington, DC

Barbier EB, Baumgartner S, Chopra K, Costello C, Duraiappah A, Hassan R, Kinzig A, Lehman M, Pascual U, Polasky S, Perrings C (2009) The valuation of ecosystem services. In: Naeem S, Bunker DE, Hector A, Loreau M, Perrings C (eds) Biodiversity ecosystem functioning and human well-being: an ecological and economic perspective. Oxford University Press, Oxford

Bennett MT (2008) China's sloping L and conversion program: institutional innovation or business as usual? Ecol Econ 65:699–711

Boyle KJ, Poor J, Taylor LO (1999) Estimating the demand for protecting the freshwater lakes from eutrophication. Am J Agric Econ 83(2):441–454

Cook A, Speldewinde P (2015) Public health perspectives on water systems and ecology. In: Finlayson CM et al (eds) Wetlands and human health. Springer, Dordrecht, pp 15–30

Cornes R, Sandler T (eds) (1996) The theory of externalities, public goods and club goods. Cambridge University Press, Cambridge

Corvalan C, Hales S, McMichael A (2005) Ecosystems and human well-being: health synthesis. World Health Organisation, Geneva

Cromie R, Davidson N, Galbraith C, Hagemeijer W, Horwitz P, Lee R, Mundkur T, Stroud DA (2011) Responding to emerging challenges: multi-lateral environmental agreements and avian influenza H5N1. J Int Wildl Law Policy 14:206–242

De Groot RS, Stuip MAM, Finlayson CM, Davidson N (2006).Valuing wetlands: guidance for valuing the benefits derived from wetland ecosystem services, Ramsar Technical Report No. 3/CBD Technical Series No. 27. Ramsar Convention Secretariat, Gland, Switzerland & Secretariat of the Convention on Biological Diversity, Montreal, Canada

Elster J (1997) The market and the forum: three varieties of political theory. In: James B, William R (eds) Deliberative democracy: essays on reason and politics. The MIT Press, Cambridge, pp 3–33

Falkenmark M, Folke C (2002) The ethics of socio-ecohydrological catchment management: towards hydrosolidarity. Hydrol Earth Syst Sci 6:1–9

Falkenmark M, Finlayson CM, Gordon LJ, Bennett EM, Chiuta TM, Coates D, Gosh N, Gopalakrishnan M, de Groot RS, Jacks G, Kendy E, Oyebande L, Moore M, Peterson JM, Portugez GD, Seesink JM, Thame K, Wasson R (2007) Agriculture, water and ecosystems: avoiding the costs of going too far. In: Molden D (ed) Water for food, water for life: a comprehensive assessment of water management in agriculture. Earthscan, London, pp 233–277

Finlayson CM, Horwitz P (2015) Wetlands as settings for human health-the benefits and the paradox. In: Finlayson CM, Horwitz P, Weinstein P (eds) Wetlands and human health. Springer, Dordrecht, pp 1–14

Finlayson CM, D'Cruz R, Davidson NJ (2005) Ecosystem services and human well-being: water and wetlands synthesis. World Resources Institute, Washington, DC

Finlayson CM, Davidson N, Pritchard D, Milton GR, MacKay H (2011) The Ramsar Convention and ecosystem-based approaches to the wise use and sustainable development of wetlands. J Int Wildl Law Policy 14:176–198

Gilbert M, Chaitaweesub P, Parakamawongsa T, Premashthira S, Tiensin T, Kalpravidh W, Wagner H, Slingenbergh J (2006) Free-grazing ducks and highly pathogenic avian influenza, Thailand. Emerg Infect Dis 12(2):227–234

Gilbert M, Xiao X, Pfeiffer DU, Epprecht M, Boles S, Czarnecki C, Chaitaweesub P, Kalpravidh W, Minh PQ, Otte MJ, Martin V, Slingenbergh J (2008) Mapping H5N1 highly pathogenic avian influenza risk in Southeast Asia. Proc Natl Acad Sci U S A 105(12):4769–4774

GRAIN (2006) Fowl play: the poultry industry's central role in the bird flu crisis. (GRAIN Briefing, February 2006). http://www.grain.org/briefings/?id=194. Accessed 19 Jan 2015

Greger M (2006) Bird flu: a virus of our own hatching. Lantern Books, New York, 465 pp

Gyrd-Hansen D (2003) Willingness to pay for a QALY. Health Econ 12:1049–1060

Harrington W, Krupnick AJ, Spofford WO Jr (1989) The economic losses of a waterborne disease outbreak. J Urban Econ 25(1):116–137

Harris-Roxas B, Harris E (2010) Differing forms, differing purposes: a typology of health impact assessment. Environ Impact Asses Rev. doi:10.1016/j.eiar.2010.03.003

Hornborg A, McNeill J, Martinez-Alier J (2007) Rethinking environmental history: world- system history and global environmental change. Altamira Press, Lanham

Horwitz P, Finlayson CM (2011) Wetlands as settings: ecosystem services and health impact assessment for wetland and water resource management. Bioscience 61:678–688

Horwitz P, Finlayson M, Weinstein P (2012) Healthy wetlands, healthy people. A review of wetlands and human health interactions. Ramsar Technical Report No. 6. Secretariat of the Ramsar Convention on Wetlands and the World Health Organization. Gland Switzerland

Hulse-Post DJ, Sturm-Ramirez KM, Humberd J, Seiler P, Govorkova EA, Krauss S, Scholtissek C, Puthavathana P, Buranathai C, Nguyen TD, Long HT, Naipospos TSP, Chen H, Ellis TM, Guan Y, Peiris JSM, Webster RG (2005) Role of domestic ducks in the propagation and biological evolution of highly pathogenic H5N1 influenza viruses in Asia. Proc Natl Acad Sci U S A 102:10682–10687

Ivey JL, Smithers J, de Loe RC, Kruetzwiser RD (2004) Community capacity for adaptation to climate-induced water shortages: linking institutional complexity and local actors. Environ Manage 33:36–47

Jardine A, Cook A, Weinstein P (2008) The utility of mosquito-borne disease as an environmental monitoring tool in tropical ecosystems. J Environ Monitor 10:1409–1414

Kallis G, Videira N, Antunes P, Guimarães Pereira, Spash CL, Coccossis H, Corral Quintana S, del Moral L, Hatzilacou D, Lobo G, Mexa A, Paneque P, Pedregal B, Santos R (2006) Participatory methods for water resource planning. Environ Plan C Gov Policy 24(2):215–234

Kavaisi A (2001) The potential for constructed wetlands for wastewater treatment and reuse in developing countries: a review. Ecol Eng 16:545–560

Kickbusch I, McCann W, Sherbon T (2008) Adelaide revisited: from healthy public policy to health in all policies. Health Promot Int 23:1–4

Krupnick AJ (2004) Valuing health outcomes: policy choices and technical issues (Technical Report). Resources for the Future, Washington, DC

Kumar R, Horwitz P, Milton G, Sellamuttu S, Buckton S, Davidson N, Pattnaik A, Zavagli M, Baker C (2011) Assessing wetland ecosystem services and poverty interlinkages: a general framework and case study. Hydrol Sci J 56(8):1602–1621

Legget CG, Bocksteal NE (2000) Evidence of the effects of water quality on residential land prices. J Environ Econ Manage 39:121–144

Mauskopf JA, French MT (1991) Estimating the value of avoiding morbidity and mortality from foodborne illnesses. Risk Anal 11:619–631

MEA (Millennium Ecosystem Assessment) (2005) Millennium ecosystem assessment synthesis report. Island, Washington, DC

Mele D (2004) The principle of subsidiarity in organisations: a case study (Working paper). IESE Business School, University of Navarra, Madrid

Miller TR (2000) Variations between countries in values of statistical life. J Transp Econ Policy 34(2):169–188

Molden D (ed) (2007) Water for food, water for life: a comprehensive assessment of water management in agriculture. Earthscan, London

Molden D, Frenken K, Barker R, de Fraiture C, Mati B, Svendsen M, Sadoff C, Finlayson CM (2007) Trends in water and agricultural development. In: Molden D (ed) Water for food, water for life: a comprehensive assessment of water management in agriculture. Earthscan, London, pp 57–89

Mooney H, Cropper A, Reid W (2005) Confronting the human dilemma: how can ecosystems provide sustainable services to benefit society? Nature 434:561–562

North D (1990) Institutions, institutional change and economic performance. Cambridge University Press, Cambridge

Ostrom E (1990) Governing the commons: the evolution of institutions for collective action. Cambridge University Press, Cambridge New York.

Patz J, Corvalan C, Horwitz P, Campbell-Lendrum D (2012) Our planet, our health, our future. Human health and the Rio conventions: biological diversity, climate change and desertification. World Health Organisation, Geneva

Perrot-Maître D (2006) The Vittel payments for ecosystem services: a "perfect" PES case? International Institute for Environment and Development, London

Rodriguez JP, Beard DT Jr et al (2006) Trade-offs across space, time and ecosystem services. Ecol Soc 11(1):28 (online)

Russi D, ten Brink P, Farmer A, Badura T, Coates D, Forster J, Kumar R, Davidson N (2013) The economics of ecosystems and biodiversity for water and wetlands. IEEP, London and Brussels, Ramsar Secretariat, Gland

Schelling T (1968) The life you save may be your own. In: Chase SB (ed) Problems in public expenditure analysis. The Brookings Institution, Washington, DC

Schoeman J Allan C, Finlayson CM (2014) A new paradigm for water? A comparative review of integrated, adaptive and ecosystem-based water management in the anthropocene. Int J Water Resour Dev 30:377–390

Shortridge KF, Melville DS (2006) Domestic poultry and migratory birds in the interspecies trans-
mission of avian influenza viruses: a view from Hong Kong. In: Boere GC, Galbraith CA,
Stroud DA (eds) Waterbirds around the world. The Stationery Office, Edinburgh, pp 427–431
Sims LD (2007) Lessons learned from Asian H5N1 outbreak control. Avian Dis 50:174–181
Sims LD, Dolmenech J, Benigno C, Kahn S, Kamata A, Lubroth J, Martin V, Roeder P (2005)
Origin and evolution of highly pathogenic H5N1 avian influenza in Asia. Vet Rec 157:159–164
Smith M, de Groot D, Bergkamp G (2006) Pay. Establishing payments for watershed services.
Gland. IUCN, Switzerland, 109 pp
Songserm T, Jam-on R, Sae-Heng N, Meemak N, Hulse-Post DJ, Sturm-Ramirez KM, Webster RJ
(2006) Domestic ducks and H5N1 influenza epidemic, Thailand. Emerg Infect Dis 12(4):575–
581
Spash CL (2007) Deliberative monetary valuation (DMV): issues in combining economic and
political processes to value environmental change. Ecol Econ 63(4):690–699
Stagl S (2007) SDRN rapid research and evidence review on emerging methods for sustainability
valuation and appraisal. (A report to the Sustainable Development Research Network). http://
www.sd-research.org.uk/wp-content/uploads/sdrnemsvareviewfinal.pdf
Starr C (1976) General philosophy of risk benefit analysis. In: Ashley H, Rudman R, Whipple C
(eds) Energy and the environment: a risk-benefit approach. Pergamon, Oxford
Stavins R (2000) Experience with market-based environmental policy instruments. In: Maler K,
Vincent J (eds) The handbook of environmental economics. Elsevier Science, Amsterdam
Stiglitz J (1986) The new development economics. World Dev 14(2):257–265
UN WWDR (2006) Water, a shared responsibility. The United Nations World Water Development
Report 2. World Water Assessment Programme, United Nations Educational, Scientific and
Cultural Organization, Paris and Berghahn Books, New York, NY
UNEP (2012) Keeping track of our changing environment: from Rio to Rio + 20 (1992–2012).
Division of Early Warning and Assessment (DEWA), United Nations Environment Programme
(UNEP), Nairobi
United Nations (2011) The millennium development goals report 2011. United Nations
United Nations Environment Program UNEP (2007) Global environment outlook 4-Environment
for Development, UNEP, Nairobi
van Eijk P, Kumar R (2009) Bio-rights in theory and practice. A financing mechanism for linking
poverty alleviation and environmental conservation. Wetlands International, Wageningen
Viscusi WK (1998) Rational risk policy. Clarendon Press
WEHAB (Water, Energy, Health, Agriculture, and Biodiversity) (2002a) A framework for action
on agriculture, World Summit on Sustainable Development, Johannesburg, South Africa
WEHAB (Water, Energy, Health, Agriculture, and Biodiversity) (2002b) A framework for action
on water, World Summit on Sustainable Development, Johannesburg, South Africa
Wilk R, Cliggett L (2006) Economies and cultures: foundations of economic anthropology,
2nd edn. Westview, Boulder
Williamson O (1985) The economic institutions of capitalism. Free, New York
Wunder S (2005) Payment for environmental services: nuts and bolts. CIFOR Occasional Paper
Zavestoski S (2004) Constructing and maintaining ecological identities: the strategies of deep
ecologists. In: Clayton S, Opotow S (eds) Identity and the natural environment: the psychologi-
cal significance of nature. The MIT Press, Cambridge, pp 297–316

Human Health and the Wise Use of Wetlands— Guidance in an International Policy Setting

C Max Finlayson and Pierre Horwitz

Abstract The Ramsar Convention has developed guidance to support the wise use of wetlands, including identifying the importance of integrating interventions aimed at biodiversity outcomes with those in support of increased human health and well-being. This guidance provides a basis for promoting the importance of wetlands as settings for human health and well-being. In doing this the Convention accepted the conceptual framework adopted for the Millennium Ecosystem Assessment and used this to assess the coverage of its wise use handbooks (guidelines or guidance on specific topics). In relation to incorporating human health and well-being into wetland management the following topics were identified as needing further attention: linking human health and well-being with wetland conservation; the maintenance of existing ecosystem services; strengthening collaboration and partnerships; development of integrated wetland policies; extending research and information sharing; assessment of the consequences of wetland management; addressing the impacts of climate change; and capacity building. The reciprocal agenda, where wetland management for ecosystem services becomes a core pursuit of the health sector, is no less extensive and no less difficult a task. The challenge remains to bring the interventions from both the health and the wetland sectors together, building on the guidance established by the Ramsar Convention for the wise use of wetlands, and incorporating their importance as settings for human health and well-being.

Keywords Wise use · Guidance · Interventions · Millennium Ecosystem Assessment · Policy · Responses · Ecosystem services · Ramsar Convention · Wetland management · Climate change

---

C. M. Finlayson (✉)
Institute for Land, Water and Society, Charles Sturt University, Albury, Australia
UNESCO-IHE, Institute for Water Education, Delft, The Netherlands
e-mail: mfinlayson@csu.edu.au

P. Horwitz
School of Natural Sciences, Edith Cowan University, Joondalup, WA, Australia
e-mail: p.horwitz@ecu.edu.au

© Springer Science+Business Media Dordrecht 2015
C. M. Finlayson et al. (eds.), *Wetlands and Human Health,* Wetlands: Ecology,
Conservation and Management 5, DOI 10.1007/978-94-017-9609-5_11

Introduction

The strong interdependence of wetland ecosystems and human health has been emphasised by Horwitz and Finlayson (2011) in order to enhance human health and well-being interventions needed to address the erosion of ecosystem services in wetlands (see also Horwitz et al. 2012 and in relation to the extent of wetland loss and degradation see MEA 2005). Suitable interventions should therefore be based around further understanding of the following: (i) linkages between the biodiversity of a wetland and ecosystems services, (ii) the manner in which ecosystem services benefit human well-being, (iii) recognition of the drivers of ecosystem change that diminish the contributions of those ecosystem services, and (iv) documentation of the outcomes of such changes. Further, all wetland management responses need to be evaluated for their possible implications for human health and well-being (Horwitz et al. 2012).

Finlayson et al. (2011) argued that the implications for human health and well-being will be influenced by the complexity of the social structures that affect wetlands and the communities that derive benefits from them. Figure 1 provides example of an array of organisations that can affect wetland management. The important point in this figure is not the specifics of the named organisations—it is the array and scope of the organisations that is important.

Figure 1 also shows how the possible response options to human health and ecosystem change might lie largely outside the direct control of both the wetland sector and the health sector. Accordingly, the importance of identifying key partners and responsible stakeholder groups required to achieve appropriate outcomes cannot be overemphasized. To be most effective, wetland policy-makers need to recognise that integrating their responses with those of partners and other stakeholders, and 'creating a space' for them, is necessary if the potential health impacts of ecosystem change are to be reduced. Hence wetland managers need to be involved in building the necessary coping capacity in human communities to ensure such outcomes are achieved, and to recognize that these responses will need to operate at local, national, or regional levels.

These concepts are not new and have, for example, been long embedded within the "wise use" approach of the Ramsar Convention on Wetlands (Finlayson et al. 2011), but they have not necessarily been articulated in a way to sufficiently embrace the wider sectors that exert an influence on wetland management (Horwitz et al. 2012). As noted, this is not a recent realisation—Hollis (1992) metaphorically pounced on this issue during the seminal conference held in 1992 to prepare a strategy to stop and reverse the loss of Mediterranean wetlands (Finlayson et al. 1992)—but the concept seems to have taken a long time to become widely accepted. This does not mean that substantive efforts have not been made to better link biodiversity with human elements in wetland management. The Ramsar Convention on Wetlands has long promoted the integration of biodiversity and human benefits from wetlands and in the late 1990s developed guidance on engaging with local communities and other sectors in order to support the wise use of wetlands and stem the loss of wetlands (Ramsar Convention Secretariat 2010). There are though ongoing

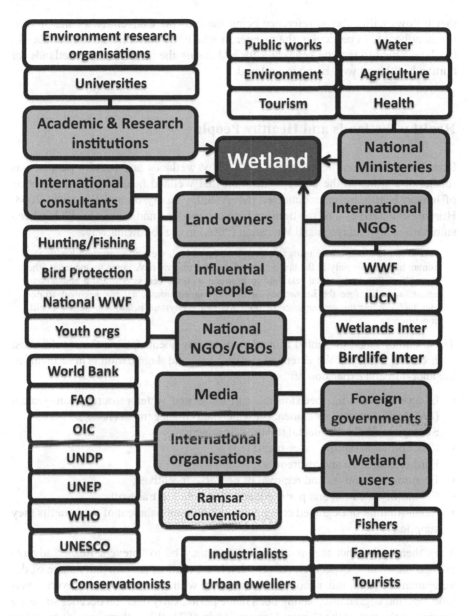

Fig. 1 An example of the array of organisations that can affect wetlands. (Adapted from Hollis 1994)

questions about how effectively Contracting Parties to the Convention have grasped and implemented such approaches (Finlayson 2012).

Despite such questions the manner in which the Ramsar Convention has addressed the wise use of wetlands provides a basis for incorporating the importance of wetlands as settings for human health and well-being. This is explored in the text that follows with specific reference being made to the Convention's suite of wise use handbooks. In particular, the manner in which the current wise use handbooks can be adapted to include key issues for addressing the importance of wetlands for human health and well-being is explored.

Healthy Wetlands and Healthy People

Recent emphasis on the aphorism of 'healthy wetlands and healthy people' has further strengthened the conceptualisation that wetland health and some aspects of human health are interconnected (MEA 2005; Finlayson and Weinstein 2008; Horwitz et al. 2012). The relationship between wetland and human health has been summarised by Finlayson and Weinstein (2008) in the following way:

> Poor wetland management leads to a deterioration of both wetland ecosystem health and human health. It is only in the last couple of decades that we have come to appreciate the strength of the fundamental relationship between wetland ecosystem health and human health, and therefore the importance of developing environmental management strategies that support the maintenance of both wetland ecosystem health and human health concurrently.

These authors then presented a number of management options that could assist with efforts to reduce the extent of wetland loss and degradation. These included the adoption and expansion of:

- Integrated planning mechanisms, e.g. integrated water resource management (IWRM), integrated catchment (or watershed) management (ICM);
- Strategic and environmental impact assessment;
- Risk and vulnerability assessment;
- Further education and awareness;
- Economic incentives and removal of perverse incentives;
- Economic valuation and participatory approaches to trade-offs;
- Rehabilitation of degraded ecosystems and re-establishment of the benefits they supplied.

They then added that the smartest response may be to "prevent further adverse change rather than expect to recover what has been lost at some later stage of development". Again, this is not a new statement with calls to stop and reverse wetland loss and degradation having been made periodically over the decades since the signing of the text of the Ramsar Convention in 1971. While there is evidence of an increase in responses to global environmental degradation these are not keeping up with the rate of degradation (Butchart et al. 2010; Armenteras and Finlayson 2011); a situation that applies to wetlands given ongoing loss and degradation (MEA 2005) and limited success in implementing the many decisions taken by Contracting Parties to the Convention (Finlayson 2012).

It is though increasingly recognised that productive, healthy wetlands with their suite of services can provide people and communities with benefits to support everyday activities and livelihoods. There are also many claims and an increasing evidence base that healthy wetlands reduce vulnerability to flood and drought, crop failure or disease, and that unhealthy wetlands can exacerbate human vulnerability by increasing the risks of flood, drought, crop failure or disease (Falkenmark et al. 2007; Gordon et al. 2011). As a consequence the concept of healthy wetlands and healthy people is being widely promoted, both popularly and scientifically (Horwitz et al. 2012), even though the term 'healthy wetlands' could do with substantial clarification (Horwitz and Finlayson 2012).

Horwitz and Finlayson (2011) reconsidered the relationship between human well-being and environmental quality, extending its relevance beyond the management of wetlands and water resources, to public health itself. They proposed an integrated strategy, involving three approaches: (i) make assessments of the ecosystem services provided by wetlands more routine; (ii) adopt the "settings" approach wherein wetlands are one of the settings for human health and provide a context for health policies; and (iii) develop a layered suite of health issues in wetland settings, including core requirements for human health (food and water), health risks from wetland exposures, and broader social determinants of health in wetland settings, including livelihoods and lifestyles. Combined, these strategies were seen as providing a way for wetland managers to incorporate health impact assessment processes into their decision making and to examine the health consequences of trade-offs that occur in wetland management, and from activities outside of their direct influence, and hence to further address the complexities of ensuring wise use of wetlands.

Wise Use of Wetlands

The imperative to conserve wetlands and provide benefits for people and for biodiversity has underpinned the substantial efforts, spread over more than a decade, to develop the Ramsar Convention on Wetlands (Matthews 1993). From the outset the text of the Convention was far reaching and contained clauses committing countries to make wise use of all wetlands and to maintain the ecological character of wetlands listed as internationally important (Ramsar sites). These two concepts were later brought together by the Convention with Contracting Parties agreeing to a commitment to maintain the ecological character of all wetlands (Pittock et al. 2010).

Unfortunately, over the same period wetlands continued to be converted and otherwise degraded with a consequent loss of benefits for people, in particular to local communities in the vicinity of wetlands (Hollis 1998; Mitsch 1998). In response to ongoing concerns about this trend, the Convention participated in the groundbreaking Millennium Ecosystem Assessment (MEA 2005) from 2000–2005 and subsequently more formally embraced the concept of ecosystem services and the benefits that accrued to people from wetlands. This included incorporating ecosystem services within the Convention's definition of ecological character and equating its wise use approach with sustainable development (Gardner and Davidson 2011; Finlayson et al. 2011). In this process the wise use of wetlands was redefined as "the maintenance of their ecological character, achieved through the implementation of ecosystem approaches, within the context of sustainable development." Ecological character was also redefined as the "combination of the ecosystem components, processes and benefits/services that characterise the wetland at a given point in time." In doing so the Contracting Parties to the Convention committed to maintaining the benefits that people derive from all wetlands—a remarkable commitment and one that could transform the management of wetlands globally.

At the same time the Convention accepted the conceptual framework adopted for the Millennium Ecosystem Assessment as a framework for the wise use of wetlands (Davidson and Finlayson 2007; Ramsar Convention Secretariat 2010). Furthermore, the Assessment provided a clear outline of how the indirect and direct drivers of change to wetland ecosystems interact and affect the capacity of a wetland to continue to deliver ecosystem services to sustain human well-being and achieve poverty reduction, and as such it was compatible with the concepts and guidance for wise use already adopted.

The framework was then used to map the contents of the Convention's wise use handbooks (guidelines or guidance on specific topics) to provide an assessment of the coverage (or gaps in it) in relation to the opportunities to address maintenance of the ecological character of all wetlands through wise use (Fig. 2). The available guidance for the wise use of wetlands is found in the widely accessible and used Handbook series produced by the Convention (now in its 4th Series, 2010).

This showed that many of the Ramsar wise use guidelines concerned strategies and interventions for wetlands and their ecological processes, or to address the direct drivers of wetland change. Further, it showed that these were focused largely at local or national levels, since Ramsar guidance is for Contracting Parties acting within their own territories, although some guidance also applied regionally and globally (e.g. see the guidance for international cooperation). The framework can also be used to illustrate how the available guidance relates to the drivers of change and possible impacts on wetland biodiversity and ecosystem services, and to identify gaps. Guidance on adaptation to climate change is one such gap, as is guidance on managing agriculture and fisheries in wetlands, both highly relevant when considering human health and well-being. These gaps are being addressed through the Convention's Scientific and Technical Review Panel (STRP), which provides independent scientific advice to the Convention.

The abovementioned changes brought the Ramsar definitions of wise use and ecological character into line with the integrated concepts for ecosystem management being promoted through the Millennium Ecosystem Assessment and overtly recognized the importance of wetland ecosystem services for human well-being and livelihoods. The interconnectedness of food and water from wetlands with human health and well-being was further explored through the Convention's involvement in the Comprehensive Assessment of Water Management in Agriculture (Molden et al. 2007; Falkenmark et al. 2007) and later engagement with the World Health Organisation (Horwitz et al. 2012).

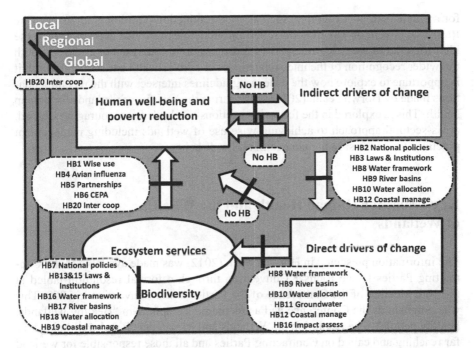

Fig. 2 Mapping of Ramsar guidance for wetland management: framework for the wise use of wetlands. (Adapted from Secretariat of the Ramsar Convention 2010). HB refers to the Ramsar wise use handbooks relevant to the guidance available for wetland management interventions at specific places in the framework. A *solid line* marks where interventions are possible

The 2009–2015 Strategic Plan of the Convention outlined and reaffirmed a range of actions and processes for Contracting Parties and others in support of the wise use of wetlands. These included: establishment of national wetland policies and plans; reviewing and harmonizing the framework of laws and financial instruments affecting wetlands; undertaking wetland inventory and assessment; integrating wetlands into sustainable development processes; ensuring public participation in wetland management and the maintenance of cultural values by local communities and indigenous people; promoting communication, education and public awareness; and increasing private sector involvement. These actions and processes intersect with those being simultaneously promoted by authors addressing the nexus between wetlands and human health and well-being.

The wise use approach of the Ramsar Convention is an example of an ecosystem approach for integrated environmental management. Ecosystem approaches were developed as an integrated alternative to sectoral approaches to environmental management and specifically for promoting conservation and sustainable use in an equitable way. They focus on managing environmental resources and human needs across landscapes and trying to balance trade-offs for both human well-being and ecosystem services and are often a response to a previous tendency to manage for a single ecosystem service, such as food production (Gordon et al. 2011). The Ramsar Convention has hitherto relied on guidelines for wise use rather than the identification of specific principles (see discussion in Finlayson et al. 2011). With the wider recognition of the inter-relatedness of human health and wetland health it is opportune to explore how the wise use guidelines intersect with the recommendations made by Horwitz et al. (2012) for ensuring the benefits of wetlands for human health. This is explored in the following sections as a basis for encouraging a broad, cross-sectoral approach to achieving wise use of wetlands including management practices for addressing human health and well-being.

Key Issues for Human Health and the Wise Use of Wetlands

The information presented in Horwitz et al. (2012) was made available to the Contracting Parties to the Ramsar Convention through a formal resolution tabled at the 10th Meeting of the Conference of the Parties to the Convention in 2008. This was poignant as the Conference had as its theme the aphorism "Healthy wetlands, healthy people". The resolution on Wetlands and human health and well-being was far reaching and called on Contracting Parties and all those responsible for wetland management to take on a series of actions and emphases as paraphrased below (and extracted from the formal text of the Resolution):

1. Link human health and well-being with wetland conservation—take action to improve the health and well-being of people in harmony with wetland conservation objectives, in particular by identifying and implementing actions that benefit both wetland ecosystems and human health concurrently;
2. Maintain existing ecosystem services—address the causes of declining human health linked with wetlands by maintaining or enhancing existing ecosystem services that can contribute to the prevention of such declines, and to ensure that any disease eradication measures in or around wetlands are undertaken in ways that do not unnecessarily jeopardise the maintenance of the ecological character of the wetlands and their ecosystem services;
3. Strengthen collaboration and partnerships—encourage all concerned to strengthen collaboration and seek new and effective partnerships between the sectors concerned with wetland conservation, water, health, food security and poverty reduction;
4. Avoid adverse impact of development activities—take all possible steps to avoid direct or indirect effects of the development sectors on wetlands that would impact negatively on those ecosystem services of wetlands that support human health and well-being;
5. Develop integrated wetland policies—make the interrelationship between wetland ecosystems and human health a key component of national and international policies, plans and strategies;

6. Extend research and information sharing—and encourage new and ongoing research regarding the links between wetlands and human health and to bring information on the scientifically proven contributions that naturally functioning wetland ecosystems make to good health and well-being to the attention of national ministries and agencies responsible for health, sanitation, and water supply;

7. Assess the consequences of wetland management—collaborate in assessing the consequences of wetland management measures linked with human health, and vice versa the consequences for the ecological character of wetlands of current practices and developments which seek to maintain or improve human health, including the identification of appropriate trade-offs in decision-making;

8. Address impacts of climate change—ensure that decision-making on co-managing wetlands and human health issues takes into account current understanding of climate change-induced increases in health and disease risk and strives to maintain the capacity of wetlands to adapt to climate change and continue to provide their ecosystem services.

9. Build capacity—dedicate resources to building capacity for more integrated approaches to wetland and water management and health, including the application of local and traditional knowledge;

10. Share information and strengthen collaboration—make available the findings of the report on "Healthy Wetlands, Healthy People" to the relevant parts of the human health community, and to discuss with the World Health Organisation ways and means of strengthening collaboration with the Ramsar Convention and the water and wetland sectors in general.

Embedding Human Health in Guidance for Wetland Management

The list above makes it clear that an organisational change is required to comprehensively address the relationship between wetlands and human health. At first sight these activities and emphases may standout as being very different from those within the existing wise use guidance which were largely (not wholly) developed in response to the ongoing loss of wetland biodiversity. However, a large number of convergent approaches and concepts already exist between the wetland and health sectors (Horwitz et al. 2012). For example, both would acknowledge the importance of greater integration between and within agencies dealing with wetlands, and both would recognise the community benefits derived from wetlands as well as community aspirations for and knowledge about wetlands.

An examination of the scope of the existing wise use handbooks suggests that there is a lot of commonality which reflects the convergence that has emerged through the recent efforts to specifically address wetlands and human health. This is shown in a comparison of the scope of the current wise use handbooks and the above listed "health activities" outlined in the resolution from the 2012 conference (found by comparing the left hand column of Table 1 to the 10 activities and emphases in the middle of the Table). This indicates that some level of guidance on wetlands and human health can already be found in the current wise use handbooks, and/or could relatively easily be located in each of the handbooks, such that the specific issues raised in the resolution on wetlands and human health can be embedded across all existing domains of activity for the Convention.

The information in the columns of Table 1 also shows that there is an absence of existing guidance on climate change and wetlands, as also shown in Fig. 2, and within the context of this paper, on the impacts of climate change on wetlands and human health. While climate change and its implications for wetland management have been considered to some extent (for example, see synthesis in Finlayson 2013) there seems to be far less direct analysis in the wetland literature of the impact on human health and well-being associated with wetlands. This is not the case when looking at the issues from a health perspective.

The adoption in the 2012 Conference of a further decision on climate change and wetlands may herald steps to address this gap given that the decision sought the development of guidance on, inter alia, the implications of climate change for maintaining the ecological character of wetlands, including the determination of appropriate reference conditions and the specification of acceptable limits of change. It is anticipated that parallel efforts by a number of organisations to develop guidance on ecosystem-based adaptation to climate change, given that such approaches incorporate human use of ecosystems, will also support the case for developing further guidance on human health and wetlands as a contribution to the wise use of wetlands.

The question for the Contracting Parties to the Convention may now move towards addressing the extent to which further guidance for wetland managers could most usefully be integrated and embedded within the existing wise use handbooks, or provided separately in a standalone form. The resolution on wetlands and human health also identified a number of further tasks for the Convention's STRP with these later being prioritised to focus on the development of guidance for the health sector on wetlands and human health, taking into account that a layer of partnership with the health sector needs to be developed with an assumption that this may provide opportunities to influence the health sector positively, and recognising that the health sector is large and heterogeneous. The third column of Table 1 summarises the full extent to which the handbooks would need revision to adequately incorporate guidance on human health issues; in many cases this would require a substantial shift in the emphasis of the guidance.

The reciprocal agenda, where wetland management for ecosystem services becomes a core pursuit of the health sector, is no less extensive and no less difficult a task. The challenge remains to bring the interventions from both the health and the wetland sectors together, building on the guidance established by the Ramsar Convention for the wise use of wetlands, and incorporating their importance as settings for human health and well-being.

Table 1 Comparison of the proposed health activities identified by the Ramsar Convention on Wetlands against the content of the existing wise use handbooks

The Ramsar Convention's wise use Handbooks	Proposed health activities are located at least partially, or can be framed, within existing Handbooks										Relevance of the wetlands and human health nexus—where guidance needs to focus more explicitly of human health and well-being
	1 Link human health and well-being with wetland conservation	2 Maintain existing ecosystem services	3 Strengthen collaboration and partnerships	4 Avoid adverse impact of development activities	5 Develop integrated wetland policies	6 Extend research and information sharing	7 Assess consequences of wetland management	8 Address impacts of climate change	9 Build capacity	10 Share information and strengthen collaboration	
Convention pillar 1: wise use											
Handbook 1—Wise use of wetlands: Concepts and approaches for the wise use of wetlands	x										Ecosystem services (benefits that humans derive for health and well-being) are part of ecological character; wise use is managing them. Introduce the concept that humans are part of wetland ecosystems, not separate to them.
Handbook 2—National Wetland Policies: Developing and implementing National Wetland Policies					x	x	x				Health sector will need to be engaged in (preferably involved in) the formulation of plans and policies

Table 1 (continued)

The Ramsar Convention's wise use Handbooks	Proposed health activities are located at least partially, or can be framed, within existing Handbooks										Relevance of the wetlands and human health nexus—where guidance needs to focus more explicitly of human health and well-being
	1 Link human health and well-being with wetland conservation	2 Maintain existing ecosystem services	3 Strengthen collaboration and partnerships	4 Avoid adverse impact of development activities	5 Develop integrated wetland policies	6 Extend research and information sharing	7 Assess consequences of wetland management	8 Address impacts of climate change	9 Build capacity	10 Share information and strengthen collaboration	
Handbook 3—Laws and institutions: Reviewing laws and institutions to promote the conservation and wise use of wetlands				x	x						Health laws (particularly those regarding sanitation, disease and environmental hazards) will need to be included and reviewed
Handbook 4—Avian Influenza and wetland: Guidance on control of and responses to highly pathogenic avian influenza	x		x								Ensure that the guidance has relevance for environmental health practitioners and health policy analysts

Table 1 (continued)

The Ramsar Convention's wise use Handbooks	Proposed health activities are located at least partially, or can be framed, within existing Handbooks										Relevance of the wetlands and human health nexus—where guidance needs to focus more explicitly of human health and well-being
	1 Link human health and well-being with wetland conservation	2 Maintain existing ecosystem services	3 Strengthen collaboration and partnerships	4 Avoid adverse impact of development activities	5 Develop integrated wetland policies	6 Extend research and information sharing	7 Assess consequences of wetland management	8 Address impacts of climate change	9 Build capacity	10 Share information and strengthen collaboration	
Handbook 5—Partnerships Key partnerships for implementation of the Ramsar Convention			x		x				x		Recognise the broad overlapping contributions that intergovernmental, governmental, and non-governmental organisations, and health-related industries can make to wetland management
Handbook 6—Wetland CEPA: The Convention's Programme on communication, education, public awareness (CEPA) 2009–2015			x			x			x		Human health outcomes and human well-being as part of the CEPA programme; recognize the community benefits derived from wetlands as well as the importance of acknowledging community aspirations for and knowledge about wetlands

Table 1 (continued)

The Ramsar Convention's wise use Handbooks	Proposed health activities are located at least partially, or can be framed, within existing Handbooks										Relevance of the wetlands and human health nexus—where guidance needs to focus more explicitly of human health and well-being
	1 Link human health and well-being with wetland conservation	2 Maintain existing ecosystem services	3 Strengthen collaboration and partnerships	4 Avoid adverse impact of development activities	5 Develop integrated wetland policies	6 Extend research and information sharing	7 Assess consequences of wetland management	8 Address impacts of climate change	9 Build capacity	10 Share information and strengthen collaboration	
Handbook 7—Participatory skills: Establishing and strengthening local communities' and indigenous people's participation in the management of wetlands			x								Recognise the direct health outcomes possible (positive or negative) when people participate in wetland management
Handbook 8—Water-related guidance: An Integrated Framework for the Convention's water-related guidance		X		x							Water quality for assessment of human health hazards; drinking water quality, sanitation

Table 1 (continued)

The Ramsar Convention's wise use Handbooks	Proposed health activities are located at least partially, or can be framed, within existing Handbooks										Relevance of the wetlands and human health nexus—where guidance needs to focus more explicitly of human health and well-being
	1 Link human health and well-being with wetland conservation	2 Maintain existing ecosystem services	3 Strengthen collaboration and partnerships	4 Avoid adverse impact of development activities	5 Develop integrated wetland policies	6 Extend research and information sharing	7 Assess consequences of wetland management	8 Address impacts of climate change	9 Build capacity	10 Share information and strengthen collaboration	
Handbook 9—River basin management: Integrating wetland conservation and wise use into river basin management		X	x	x	x				x	x	River basins as the 'settings' for human health, livelihoods and human well-being
Handbook 10—Water allocation and management: Guidelines for the allocation and management of water for maintaining the ecological functions of wetlands		X									Ensuring sufficient water or appropriate water regimes for humans not to be exposed to human health hazards

Table 1 (continued)

The Ramsar Convention's wise use Handbooks	Proposed health activities are located at least partially, or can be framed, within existing Handbooks										Relevance of the wetlands and human health nexus—where guidance needs to focus more explicitly of human health and well-being
	1 Link human health and well-being with wetland conservation	2 Maintain existing ecosystem services	3 Strengthen collaboration and partnerships	4 Avoid adverse impact of development activities	5 Develop integrated wetland policies	6 Extend research and information sharing	7 Assess consequences of wetland management	8 Address impacts of climate change	9 Build capacity	10 Share information and strengthen collaboration	
Handbook 11—Managing groundwater: Managing groundwater to maintain ecological character		X									Managing groundwater to ensure humans are not exposed to chemical or biological hazards
Handbook 12—Coastal management: Wetland issues in Integrated Coastal Zone Management		X	x	x					x	x	Managing coastal zones to ensure humans are not exposed to physical or biological hazards

Table 1 (continued)

The Ramsar Convention's wise use Handbooks	Proposed health activities are located at least partially, or can be framed, within existing Handbooks										Relevance of the wetlands and human health nexus—where guidance needs to focus more explicitly of human health and well-being
	1 Link human health and well-being with wetland conservation	2 Maintain existing ecosystem services	3 Strengthen collaboration and partnerships	4 Avoid adverse impact of development activities	5 Develop integrated wetland policies	6 Extend research and information sharing	7 Assess consequences of wetland management	8 Address impacts of climate change	9 Build capacity	10 Share information and strengthen collaboration	
Handbook 13—Inventory, assessment, and monitoring: An Integrated Framework for wetland inventory, assessment, and monitoring		X					x				Assessment and monitoring for human health; requires development of indicators for the relationship between wetland management and human health outcomes
Handbook 14—Data and information needs: A Framework for Ramsar data and information needs						x	x		x	x	Health data and information required for health impact assessments for wetland management actions, and burden of disease assessments for wetlands themselves.

Table 1 (continued)

The Ramsar Convention's wise use Handbooks	Proposed health activities are located at least partially, or can be framed, within existing Handbooks										Relevance of the wetlands and human health nexus—where guidance needs to focus more explicitly of human health and well-being
	1 Link human health and well-being with wetland conservation	2 Maintain existing ecosystem services	3 Strengthen collaboration and partnerships	4 Avoid adverse impact of development activities	5 Develop integrated wetland policies	6 Extend research and information sharing	7 Assess consequences of wetland management	8 Address impacts of climate change	9 Build capacity	10 Share information and strengthen collaboration	
Handbook 15—Wetland inventory: A Ramsar Framework for wetland inventory						x	x				Framework to include health criteria; i.e. where wetlands are explicitly designed or used as places that improve health and well-being, perhaps through recreational use
Handbook 16—Impact assessment: Guidelines for incorporating biodiversity-related issues into environmental impact assessment legislation and/or processes and in strategic environmental assessment				x			x				Incorporate Health Impact Assessment into wetland management planning

Table 1 (continued)

The Ramsar Convention's wise use Handbooks	Proposed health activities are located at least partially, or can be framed, within existing Handbooks										Relevance of the wetlands and human health nexus—where guidance needs to focus more explicitly of human health and well-being
	1 Link human health and well-being with wetland conservation	2 Maintain existing ecosystem services	3 Strengthen collaboration and partnerships	4 Avoid adverse impact of development activities	5 Develop integrated wetland policies	6 Extend research and information sharing	7 Assess consequences of wetland management	8 Address impacts of climate change	9 Build capacity	10 Share information and strengthen collaboration	
Convention pillar 2: Ramsar sites designation and management											
Handbook 17—Designating Ramsar Sites: Strategic Framework and guidelines for the future development of the List of Wetlands of International Importance		X									Designate Ramsar sites for their ecosystem services that contribute significantly to human health and well-being
Handbook 18—Managing wetlands: Frameworks for managing Ramsar sites and other wetlands	x			x		x	x				Include human health issues in wetland management planning

Table 1 (continued)

The Ramsar Convention's wise use Handbooks	Proposed health activities are located at least partially, or can be framed, within existing Handbooks										Relevance of the wetlands and human health nexus—where guidance needs to focus more explicitly of human health and well-being
	1 Link human health and well-being with wetland conservation	2 Maintain existing ecosystem services	3 Strengthen collaboration and partnerships	4 Avoid adverse impact of development activities	5 Develop integrated wetland policies	6 Extend research and information sharing	7 Assess consequences of wetland management	8 Address impacts of climate change	9 Build capacity	10 Share information and strengthen collaboration	
Handbook 19—Addressing change in wetland ecological character Addressing change in the ecological character of Ramsar Sites and other wetlands	x			x			x				Change should include where wetlands have become the source of hazards to human health; or, loss of benefits derived from wetlands for human health, livelihoods or well-being

Table 1 (continued)

The Ramsar Convention's wise use Handbooks	Proposed health activities are located at least partially, or can be framed, within existing Handbooks										Relevance of the wetlands and human health nexus—where guidance needs to focus more explicitly of human health and well-being
	1 Link human health and well-being with wetland conservation	2 Maintain existing ecosystem services	3 Strengthen collaboration and partnerships	4 Avoid adverse impact of development activities	5 Develop integrated wetland policies	6 Extend research and information sharing	7 Assess consequences of wetland management	8 Address impacts of climate change	9 Build capacity	10 Share information and strengthen collaboration	
Convention pillar 3: international cooperation											
Handbook 20—International cooperation: Guidelines for international cooperation under the Ramsar Convention on Wetlands							x			x	International cooperation should include Conventions and organizations whose focus is on human health. Ramsar also can take a lead with biodiversity and human health issues.

Table 1 (continued)

The Ramsar Convention's wise use Handbooks	Proposed health activities are located at least partially, or can be framed, within existing Handbooks										Relevance of the wetlands and human health nexus—where guidance needs to focus more explicitly of human health and well-being
	1 Link human health and well-being with wetland conservation	2 Maintain existing ecosystem services	3 Strengthen collaboration and partnerships	4 Avoid adverse impact of development activities	5 Develop integrated wetland policies	6 Extend research and information sharing	7 Assess consequences of wetland management	8 Address impacts of climate change	9 Build capacity	10 Share information and strengthen collaboration	
Companion document											
Handbook 21—The Ramsar Convention Strategic Plan 2009–2015 Goals, strategies, and expectations for the Ramsar Convention's implementation for the period 2009–2015	X	X	x	x	x	x	x		x	x	Recognise health and well-being across all fields and activities as a strategic cross-cutting theme

References

Armenteras D, Finlayson CM (2011) Biodiversity. In: UNEP (ed) Keeping track of our changing environment: from Rio to Rio+20 (1992-2012). Division of Early Warning and Assessment (DEWA). United Nations Environment Programme (UNEP), Nairobi, pp 133–166

Butchart SHM, Walpole M, Collen B, van Strien A, Scharlemann JPW, Almond A, Baillie JE, Bomhard B, Brown C, Bruno J, Carpenter KE, Carr GM, Chanson J, Chenery AM, Csirke J, Davidson NC, Dentener F, Foster M, Galli A, Galloway JN, Genovesi P, Gregory RD, Hockings M, Kapos V, Lamarque J-F, Leverington F, Loh J, McGeoch MA, McRae L, Minasyan A, Hernandez Morcillo M, Oldfield TEE, Pauly D, Quader S, Revenga C, Sauer JR, Skolnik B, Spear D, Stanwell-Smith D, Stuart SN, Symes A, Tierney M, Tyrrell TD, Vie J-C, Watson R (2010) Global biodiversity: indicators of recent declines. Science 328(5892):1164–1168

Davidson NC, Finlayson CM (2007) Developing tools for wetland management: inventory, assessment and monitoring-gaps and the application of satellite-based radar. Aquat Conserv Marine Freshw Ecosyst 17:219–228

Falkenmark M, Finlayson CM, Gordon L (coordinating lead authors) (2007) Agriculture, water, and ecosystems: avoiding the costs of going too far. In: Molden D (ed) Water for food, water for life: a comprehensive assessment of water management in agriculture. Earthscan, London, pp 234–277

Finlayson CM (2012) Forty years of wetland conservation and wise use. Aquat Conserv Marine Freshw Ecosyst 22:139–143

Finlayson CM, Weinstein P (2008) Wetlands, health and sustainable development-global challenges and opportunities. In: Ounsted M, Madgwick J (eds) Healthy wetlands, healthy people. Wetlands International, Wageningen, pp 23–40

Finlayson CM, Hollis GE, Davis TJ (eds) (1992) Managing Mediterranean wetlands and their birds. IWRB Special Publication No. 20, Slimbridge, UK, 285 p

Finlayson CM, Davidson N, Pritchard D, Milton GR, MacKay H (2011) The Ramsar Convention and ecosystem-based approaches to the wise use and sustainable development of wetlands. J Int Wildl Law Policy 14:176–198

Gardner RC, Davidson NC (2011) The Ramsar Convention. In: LePage BA (ed) Wetlands: integrating multidisciplinary concepts. Springer, Dordrecht, pp 189–203

Hollis GE (1992) The causes of wetland loss and degradation in the Mediterranean. In: Finlayson CM, Hollis GE, Davis TJ (eds) Managing Mediterranean wetlands and their birds. IWRB Special Publication No. 20, Slimbridge, UK, pp 83–90

Hollis GE (1994) Halting and reversing wetland loss and degradation: a geographical perspective on hydrology and land use. Thomas Telford Services and Institution of Engineers, London

Hollis GE (1998) Future wetlands in a world short of water. In: McComb AJ, Davis JA (eds) Wetlands for the future. Gleneagles Publishing, Adelaide, pp 5–18

Horwitz P, Finlayson CM (2011) Wetlands as settings: ecosystem services and health impact assessment for wetland and water resource management. Bioscience 61:678–688

Horwitz P, Finlayson M, Weinstein P (2012) Healthy wetlands, healthy people: a review of wetlands and human health interactions. Ramsar Technical Report w 5 No. 6. Secretariat of the Ramsar Convention on Wetlands, Gland, Switzerland, & The World Health Organization, Geneva, Switzerland

Matthew GVT (1993) The Ramsar Convention on wetlands: its history and development. Ramsar Convention Bureau, Gland

MEA (Millennium Ecosystem Assessment) (2005) Ecosystems and human well-being: synthesis. Island Press, Washington

Mitsch WJ (1998) Predicting the world's wetlands: threats and opportunities in the 21st Century. In: McComb AJ, Davis JA (eds) Wetlands for the future. Gleneagles Publishing, Adelaide, pp 19–32

Molden D, Frenken K, Barker R, de Fraiture C, Mati B, Svendsen M, Sadoff C, Finlayson CM (2007) Trends in water and agricultural development. In: Molden D (ed) Water for food, water for life: a comprehensive assessment of water management in agriculture. Earthscan, London, pp 57–89

Pittock J, Finlayson CM, Gardner A, McKay C (2010) Changing character: the Ramsar Convention on wetlands and climate change in the Murray-Darling basin, Australia. Environ Plan Law J 27:401–442

Ramsar Convention Secretariat (2010) Wise use of wetlands: concepts and approaches for the wise use of wetlands. Ramsar handbooks for the wise use of wetlands, 4th edn, vol 1. Ramsar Convention Secretariat, Gland

A Synthesis: Wetlands as Settings for Human Health

C Max Finlayson, Pierre Horwitz and Philip Weinstein

Abstract The interactions between wetlands and people have been explored in this book through the treatment of ecological and human health and well-being issues for both wetland management and public health practitioners. This recognises that both sectors have reciprocal and important roles to play in ensuring that the benefits provided by wetlands are maintained and even enhanced. Examples are given of the benefits for human health and well-being derived from wetlands, as well as the potential for adverse outcomes if the ecological character of wetlands is not maintained when making decisions about wetlands and human health issues. The examples provide more resolution to what it means for a wetland to be 'healthy' with a proposition that wetland health should be based on social values and indicators that could be agreed through the following steps: establish the best possible reference condition, given acceptable land or water use; make judgements based on uses of human amenity derived from the wetland; acknowledge that restoration may be necessary, especially where wetland uses prove to be non-sustainable; and accept that changes in use/amenity can change the condition and hence perception of the health of the wetland.

The 'settings' approach for wetlands is promoted whereby the wetland is the 'setting' in which people "*take care of each other, our communities and our natural environment*". The setting also includes the institutional and governmental aspects required to deliver health services, to address health inequalities, and to intervene for public health. The key message from this book is that wetlands (as places of water on land, and where water shapes the land), and human health (which in its richest sense addresses the well-being of people, beyond

C. M. Finlayson (✉)
Institute for Land, Water and Society, Charles Sturt University, Albury, Australia
UNESCO-IHE, Institute for Water Education, Delft, The Netherlands
e-mail: mfinlayson@csu.edu.au

P. Horwitz
School of Natural Sciences, Edith Cowan University, Joondalup, WA, Australia
e-mail: p.horwitz@ecu.edu.au

P. Weinstein
School of Biological Sciences, The University of Adelaide, Adelaide 5005, Australia
e-mail: philip.weinstein@adelaide.edu.au

© Springer Science+Business Media Dordrecht 2015
C. M. Finlayson et al. (eds.), *Wetlands and Human Health*, Wetlands: Ecology, Conservation and Management 5, DOI 10.1007/978-94-017-9609-5_12

ill-health or the absence of disease), are interconnected and to a certain extent interdependent. Maintaining or restoring wetlands can promote the ecosystem services that support the many benefits that provide vital support for human health and well-being.

Keywords Wetland settings · Ramsar convention · Wise use · Disease · Public health · Food pollution · Livelihoods · Disasters · Guidance · Intervention · Management

Introduction

The interactions between wetlands and people have been explored in this book through a general treatment of ecological and human health and well-being issues for both wetland management and public health practitioners. This recognises that both sectors, and the disciplines that support them, have reciprocal and important roles to play in ensuring that the benefits provided by wetlands are maintained and even enhanced. At the same time there is substantial evidence to demonstrate where wetlands have been lost and degraded resulting in adverse health and well-being outcomes for local communities (see global assessment presented in MEA 2005).

Another perspective of wetland ecosystems is where there is a complex interplay between the health and well-being of components of wetland ecosystems (including humans), and ecosystem processes and functions. As an example of this shift in emphasis to one of humans being considered as part of ecosystems rather than the singular threat to them, the Ramsar Convention on Wetlands seeks to place more emphasis on understanding how people and wetlands interact, including the interactions that occur between human health and wetlands (Horwitz et al. 2012). This emphasis provides a connection with the wise use of wetlands approach promoted through the Ramsar Convention (Finlayson et al. 2011) which provides a basis for ensuring the maintenance of the ecosystem services that support human health and well-being. These interactions have been loosely encompassed within the metaphor 'healthy wetlands, healthy people' that conveys an intimate relationship between wetlands and people (Finlayson and Weinstein 2008; Horwitz et al. 2012) and supports the application of interdisciplinary approaches that have been the hallmark of recent global assessments that have examined human well-being and ecosystem services (Finlayson and Horwitz 2015a).

Information on the interactions between human health and wetlands has been provided in this book through the 'lens' of the ecosystem services available in and from wetlands. In doing so it has been implicit that human health should be treated as a component of human well-being that is itself inextricably linked with broader systemic conditions found within wetlands, and that this set of relationships can be referred to as "wetland health". It follows that wetland management has a mandate to attend to these dimensions.

Locating Human Health in the Context of Wetlands

Where wetlands are involved in the lived experiences of local people, there are outcomes for their health and well-being. Horwitz et al. (2012) emphasised that wetlands could "… *either enhance or diminish human health depending on the ecological functioning of wetlands and their ability to provide ecosystem services.*". At the extremes there are situations where people and wetlands both benefit (the double dividend), and where a detriment might be felt for both (a double negative). But the simple argument that healthy wetlands mean healthy people may not sufficiently recognise the complexity of the linkages that exist between people and wetlands nor the social and environmental changes that influence these linkages (Finlayson and Weinstein 2008). A more realistic interpretation is that wetland health incorporates the health of its components, and humans are one of those components, and that the interactions between people and wetlands are complex and multi-faceted and associated with local circumstances and drivers of change (MEA 2005; Finlayson and Weinstein 2008; Horwitz et al. 2012).

By perceiving people as part of ecosystems rather than outside of them, in this case belonging to wetland ecosystems, the meanings of the double dividend and the double negative outcomes become unmistakeable. But paradoxical outcomes also become clear: in some circumstances well-managed wetlands even those that have been hydrologically unaltered or functionally maintained, can provide poor outcomes for human health; and the converse, that human health can be supported even when wetlands are degraded (though for how long no-one is quite sure; see Raudesepp-Hearne et al. 2010).

With this background, the purpose of this book was to review and map the relationships and issues concerning human health in the context of the wise use of wetland ecosystems, including specific issues associated with the benefits derived from ecosystem services and the paradoxes that also occur.

Linkages Between Human Health and Wetlands

In order to address these issues and consider the trade-offs that may occur between wetland health and the multiple ways in which wetlands are managed (or mismanaged), with subsequent impacts on human health and well-being, the following topics have been covered in individual chapters:

- Public health perspectives (Cook and Speldewinde 2015)
- Basic needs, food security and medicinal properties (Cunningham 2015)
- Sites of exposure to water-borne infectious diseases (Derne et al. 2015)
- Human exposures to pollutants and toxicants in wetlands (Horwitz and Roiko 2015)
- Mosquito borne disease (Carver et al. 2015)
- Livelihoods and human health (McCartney et al. 2015)
- Urban wetlands and community well-being (Carter 2015)

- Natural disasters, health and wetland (Jenkins and Jupiter 2015)
- International guidance for wise use of wetlands (Finlayson and Horwitz 2015b)
- Interventions for enhancing human well-being (Horwitz et al. 2015)

The key issues raised in the individual chapters are presented below.

Public Health Perspectives

Cook and Speldewinde (2015) provide a public health perspective on wetlands with a summary of the relationship between public (community) health and the water cycle and how water systems and their ecology relate to public health, and emphasise the value of systematic approaches for assessing the health risks that can arise from water contaminants. This includes an examination of how two important public health activities—the application of epidemiological methods and the use of systems of surveillance—relate to water sources and supplies and recognises that biological and chemical contamination can originate from or be mitigated by wetlands. Hence, investigations into public health risks need to focus on potential exposure pathways and consider how wetlands or specific features within wetlands are linked to adverse health effects. This further implies that an understanding of the features of wetlands that could affect public health could assist in epidemiological investigations and surveillance programs.

The importance of wetlands for public health are realised when the many potential consequences for public health from exposure to wetlands are considered. These include those that affect hydration and potable water, nutrition, exposure to pollutants or toxicants, or to infectious diseases, and physical hazards, as well as being settings for livelihoods and mental health and psycho-social well-being, and sites from which medicinal and other products can be derived. As many wetlands are under threat from development activities there is increasing recognition that degradation could affect human health and well-being, for example through changes to wetlands that cause an increase in exposure to vector borne diseases or the loss of a protein source for local communities.

Given the extent of wetland degradation, and the relationship between human health and well-being and the condition of wetland ecosystems, public health authorities need to be aware of the potential for the emergence or re-emergence of wetland-linked diseases, and where necessary to act preventatively and proactively when developing responses. As the ecological outcomes from wetland degradation as well as from efforts to restore wetlands are largely uncertain there could be unexpected consequences for human health.

Basic Needs, Food Security and Medicinal Properties

Cunningham (2015) describes the contributions wetlands make to food security for local people, in particular through the provision of high water quality, protein, and edible or medicinal plants, as well as a means of obtaining income by trading resources, including fish, shell-fish or fibrous plants. There is also a sustained interest in seeking traditional medicines and new natural products from wetland environments such as hot springs, alpine wetlands, particularly in high diversity montane systems, desert salt-pans, soda lakes, highly alkaline or acid streams and high diversity tropical rivers.

The importance of the benefits derived from wetland products can mediate some of the negative aspects associated with wetlands and generate support for policy reform and action for wetland conservation. Understanding the links between ecosystem services and human health, especially for local and indigenous communities with close associations with wetlands, provides a platform for addressing changes in policy and practices to ensure the benefits are maintained or restored. The difficulties of making such changes are recognized, especially when attempted at a catchment scale with attendant complexity and competing needs, necessitating trade-offs, co-management and/or incentive schemes, including payment for ecosystem services to ensure some of the wider benefits can be sustained. Schemes for payment for ecosystem services have been developed, but are still at an early stage. Nevertheless, they are seen as an important option for policy makers and communities alike.

Sites of Exposure to Water-Borne Infectious Diseases

Derne et al. (2015) provide an overview of the diversity of infectious water-borne diseases that humans are exposed to as a result of their association with wetlands, and of the manner in which environmental and human factors interact to determine the overall risk of infection. A proportion of the organisms found in wetlands can cause diseases in humans with an expected increase in the disease burden as pressure builds on wetlands, particularly where sanitation infrastructure is poor. Understanding the interactions between the multiple factors affecting such diseases is difficult and often situation-specific as almost all outbreaks or emergence of disease result from a unique set of circumstances. This necessitates a multidisciplinary approach in order to establish the potential health gain from interventions that are aimed at improving ecosystem and human health concurrently, much like the 'One Health' concept where human health is recognised as being intricately linked with that of animals, and with the environment in which they coexist.

Knowledge and methods from multiple disciplines are needed to create integrated management plans for wetlands as settings for wetland conservation and for minimising the disease burden faced by humans. This includes acknowledging the importance of risk mitigation steps that involve the prevention of contamination, providing adequate sanitation and maintaining or restoring healthy wetlands. Future research needs to include investigations of the complexities of wetland ecosystems and the interlinked factors that affect the risk of infection.

Human Exposures to Pollutants and Toxicants in Wetlands

Horwitz and Roiko (2015) consider the exposure of humans to pollutants and toxicants in wetland settings and identify two principal forms of human exposures: (i) where the exposure is determined by the service that is provided (for example the presence of a contaminant in drinking water), and (ii) where the conditions for exposure occur when services are eroded (for example where an oversupply of nutrients overwhelms the water purification capacities of wetlands resulting in an exposure to a microbial toxin). Human activities greatly affect the nature of such exposures, especially where pollution is involved, with complex interactions and difficulties for environmental health practitioners. While steps can be taken to minimise the health risks resulting from such exposures, the risks can increase if the wetlands and their ecosystem services are disrupted. For any resultant interventions to be effective the imbalance in ecosystem services needs to be addressed.

An ecosystem approach provides an alternative to the more traditional risk assessment/risk management approach for addressing health risks from environmental chemicals. This includes determining the upstream socio-cultural and political causes of toxicant accumulation and exposure, and examining the wetland settings and the trade-offs that are made between ecosystem services. Ecosystem restoration, including ecosystem services, could address many of the upstream factors responsible for hazardous environmental exposures to pollutants and other toxicants by humans, but are rarely considered within the scope of public health interventions.

Mosquito Borne Disease

Carver et al. (2015) review and evaluate the links that occur between mosquitoes that breed in wetlands and disease transmission, and the natural mechanisms that regulate these. A particular focus is how wetland health can influence the transmission of disease given the importance of mosquitoes as vectors of pathogens to wildlife, livestock and humans. Despite a paucity of information there is a general belief that healthy wetlands minimise the risk of mosquito-borne disease to surrounding human and animal populations by regulating the production and dominance of mosquito species that carry or host such diseases. Given the importance of these diseases, the lack of data is surprising and suggests that the management of such diseases and their causes is reactive, including a predisposition towards the use of chemicals to control mosquito populations. It may also reflect the absence of any meaningful connection between the public health sector responsible for managing infectious disease transmission and those responsible for managing wetlands. This is an important gap as the interactions between mosquitoes, invertebrates, and vertebrates in wetlands are complex and greatly affected by human activities that can degrade wetlands or disrupt the balances that would have occurred naturally.

In a public health paradigm, the natural ecological interactions that occur in wetlands can be considered a direct ecosystem service, namely the natural mitigation of vector-borne disease risk. This is an important service given the impact of some diseases such as malaria. The disruption of these interactions and the balances that occur in wetlands due to land-use activities, habitat alteration and biodiversity loss, and climatic changes, could disrupt the ecological processes that regulate mosquito populations and have severe implications for human health. The maintenance of healthy wetlands is likely to have benefits for human health, and provide a more cost effective and sustainable way forward than chemical control of vector species. Future multidisciplinary research that can bring together the processes that regulate the transmission of mosquito-borne diseases is likely to have significant consequences for human health, as well as provide a dividend for biodiversity through the maintenance of healthy wetlands.

Livelihoods and Human Health

McCartney et al. (2015) consider the importance of wetlands for supporting and sustaining the livelihoods and hence the health of the millions of people who depend on subsistence agriculture, at times with limited access to basic human needs such as food and water. Under these conditions, wetlands, through the provision of a range of ecosystem services, are a vital asset because of their contribution to basic human needs. The "natural capital" of wetlands can be transformed, either directly or indirectly, to other forms of capital that can support their livelihoods and well-being.

Throughout much of the developing world wetlands are places where many people live and are vectors for wherefrom they derive much of their livelihoods. The extent to which people depend on wetlands is highly specific and influenced by the diversity of wetlands as well as local social-economic and cultural circumstances.

For many people, wetlands underpin their food and water security as well as other tangible and intangible benefits that affect their health, including their safety. Some of these benefits are derived directly from wetlands, such as food and medicinal plants, but, in common with all forms of natural capital, many others are only realised when the natural capital supplied by wetlands is switched to other forms of livelihood capital, through trade and commerce. However, wetlands can also harbour pests and disease (for example, by providing breeding habitat for mosquitoes that transmit malaria and snails that are vectors for schistosomiasis) that can undermine the health and livelihoods of many people. From a human perspective the benefits and detriments conferred by wetlands vary considerably with vast differences between wetland types, within a wetland type and even, spatially and temporally, within a single wetland. These differences need to be considered when developing strategies to support efforts to reduce the poverty of local communities associated with wetlands, particularly when they depend on the benefits provided by wetlands. Economic development that degrades and undermines the productivity and sustainability of wetlands could just as readily undermine the natural capital on which the poorest and most vulnerable people depend. While support to improve the livelihoods of local people is not necessarily directly congruent with conservation objectives, there can be significant and adverse livelihood and health outcomes if the balance between conservation and development is not adequately considered.

Urban Wetlands and Community Well-being

Carter (2015) explores the positive health benefits of cultural ecosystem services, including improved physical and psychological health, increased community connection and sense of place, and those derived from community involvement in urban conservation, associated with the use and enhancement of urban wetlands. These benefits are seen as a counterbalance to the detrimental interactions between wetlands and people as a consequence of wetland degradation and exposure to toxicants. The benefits are derived from the role that wetlands play in improving the quality of human surroundings and providing aesthetically pleasing places for recreation, education and spiritual development.

Cultural services, such as those mentioned above, have positive relationships to human health and are at the centre of efforts to prevent further loss and degradation of urban wetlands and where, possible, to restore them. For conservation and restoration to be successful the values of cultural services from urban wetlands need to be clearly articulated and used to guide decisions about urban planning. Thoughtful decisions about the placement of urban wetlands and the benefits available to urban communities could engage and empower people to visit and care for them. The latter activity could be supported through plans to survey and evaluate the services and the level of community benefits that are provided.

Natural Disasters, Health and Wetland

Jenkins and Jupiter (2015) review the direct and indirect health consequences of interruptions to wetland ecosystem services associated with disaster events in the context of small island developing states in the Pacific and emphasize how longer-term health effects of natural disasters can be exacerbated when wetland services are lost. Given their geographic isolation and level of economic development these states are seen as sensitive to natural disasters and may lack the adaptive capacity to effectively respond especially as the threat to public health from natural disasters may be growing. Additionally, the effects of natural disasters will interact with effects of climate variability and rapid environmental change.

Because the extent to which wetlands can mediate the impact of disasters and assist recovery is not well understood, it is recommended that the role of wetlands and disaster-related epidemiology be further investigated. Doing this could strengthen existing models for disaster risk management and wetland conservation, taking into account that wetlands can mitigate or contribute to health outcomes from disasters. In this respect the health risks from disasters that affect people living in the vicinity of wetlands or are dependent on wetland services cannot be considered in isolation of the wetland setting. The importance of considering the connections between wetland settings and disasters is accompanied by a note of caution about the need for specific information and evaluation to ensure the connections are understood and included, when appropriate, in disaster recovery activities. This could include measures to protect and restore wetlands, such as mangroves and those in riparian zones, that can help mitigate human exposure to water pollution and storm surge associated with disasters. It could also support measures to maintain ecosystem services, such as fisheries, that contribute to livelihoods, and the cultural (including psycho-social) factors that contribute to human well-being.

International Guidance for Wise Use of Wetlands

Finlayson and Horwitz (2015b) review the guidance developed by the Ramsar Convention to support the wise use of wetlands as a basis for promoting the importance of wetlands as settings for human health and well-being. The following topics were identified as needing further attention: linking human health and well-being with wetland conservation; the maintenance of existing ecosystem services; strengthening collaboration and partnerships; development of integrated wetland policies; extending research and information sharing; assessment of the consequences of wetland management; addressing the impacts of climate change; and capacity building.

The reciprocal agenda, where wetland management for ecosystem services becomes a core pursuit of the health sector, is no less extensive and no less difficult to achieve. The challenge remains to bring the interventions from both the health and the wetland sectors together. For the wetland sector this could readily build on the extensive guidance provided by the Ramsar Convention for the wise use of wetlands, and incorporate the importance of wetlands as settings for human health and well-being. Reciprocal guidance for the health sector would strengthen the collaboration and actions needed to ensure that the importance of wetlands for human health and well-being is more widely considered. The reciprocal relationship was partly achieved by the Convention's response to the Millennium Ecosystem Assessment whereby purposeful steps were taken to embed human health in guidance for wetland management. The development of guidance for the health sector on wetlands and human health is seen as an important further step, recognising that the sector is both large and diverse.

The convergence that has emerged through recent efforts to specifically address wetlands and human health is evident through the scope of the current 'wise use' handbooks provided by the Convention. While the wise use handbooks were largely produced to support wetland conservation from a biodiversity viewpoint, they do contain guidance that can be used to support the reciprocal relationship between wetlands and human health. This represents a convergence in the conceptual approaches that have underpinned the wetland and health sectors, recalling that the Convention has from the outset recognised the importance of wetlands for people.

There is an absence of existing guidance on climate change and wetlands, and the impacts of climate change on wetlands and human health. Widely dispersed efforts to develop guidance on ecosystem-based adaptation to climate change are expected to support the case for developing further guidance on human health and wetlands.

Interventions for Enhancing Human Well-Being

Horwitz et al. (2015) recognise that many of the possible response options for addressing change in wetlands and human health and well-being are outside the direct control of the wetland sector, or even the health sector. Instead they are embedded elsewhere, including with authorities responsible for sanitation and water supply, education, agriculture, trade, tourism, transport, development, and housing. As a consequence, inter-sectoral and cross-sectoral integrated options or interventions are needed if the potential impacts of wetland degradation on human health and well-being are to be mitigated or avoided.

Interventions will need to take into account existing social values and cultural norms, existing infrastructure, and the social, economic, demographic, and political driving forces that result in wetland change. Key components of such interventions will include (i) engagement with representatives from previously marginalised stakeholders, (ii) increased transparency and exchange of information between and within sectors, and (iii) recognition of the core pursuits of other sectors. The interventions will vary enormously given local circumstances and range from (i) promoting cross-sectoral governance and institutional structures, (ii) promoting rationalized incentive structures, (iii) social and behavioural responses which include capacity building, communication and empowerment; and (iv) technological solutions to enhance the multi-functionality of ecosystems. As some interventions may involve tradeoffs between the benefits derived from wetlands, and between stakeholders, it is important to understand the consequences of choosing one option in preference to another. While recognising the potential for tradeoffs is an important step it will also be necessary to establish transparent processes by which these can be negotiated especially given the close relationship that exists between food production, water use and water extraction. The closeness of this relationship drives a broad societal objective for those charged with wetland management, an objective that extends far beyond nature conservation and a simplistic approach to natural resource management. Because the pressures that impair the capacity of many communities to prepare for their future are very much the same as those that can impede attempts to make use of wetlands, it is necessary for wetland managers to work with communities if the human health and well-being benefits of healthy wetlands are to be realised. This can include capacity building to develop community resilience in support of wider efforts to help achieve the Millennium Development Goals, and the next generation thereof post 2015.

Healthy Wetlands: On Settings and Services

The above text provides examples of both the benefits for human health and well-being derived from wetlands, as well as the potential for adverse outcomes if the ecological characteristics of wetlands are not considered when making decisions about wetlands and human health issues. This includes the potential for livelihoods to be adversely affected through the loss of access to basic needs such as clean water and food, as well as increased exposure to disease, or to physical hazards such as flooding. The importance of the ecosystem services derived from wetlands has been widely illustrated and documented since the publication of the landmark reports by the Millennium Ecosystem Assessment (MEA 2005) and the Ramsar Convention (Horwitz et al. 2012) on human health, well-being and wetlands.

The perspectives and the examples given in this book provide more resolution to what it means for a wetland to be 'healthy'. Finlayson and Weinstein (2008) proposed that wetland health should be based on social values and indicators and that this would be consistent with ecosystems services as a component of the eco-logical character of wetlands. They argued that a socially-oriented approach could comprise the following steps:

- establish the best possible reference condition, given acceptable land or water use;
- make judgements based on uses of human amenity derived from the wetland;
- acknowledge that restoration may be necessary, especially where wetland uses prove to be non-sustainable; and
- accept that changes in use/amenity can change the condition and hence percep-tion of the health of the wetland.

Horwitz and Finlayson (2011) argued that "A claim to "healthy ecosystems" comes from the inclusion of the systems thinking required to make judgments on the desir-ability of an ecological character. It is also explicit about the health of components of the ecosystem (including humans), and whether organizations are adaptive and responsive to ecosystem changes."

Both definitions draw on the ecosystem as a location, providing goods and ser-vices, and a context for the way people live their lives and make decisions. This is consistent with the dual concepts of 'settings' and 'ecosystem services'.

The 'settings' approach from health promotion was first enunciated in the Otta-wa Charter (WHO 1986) whereby the wetland is the 'setting' in which people "*take care of each other, our communities and our natural environment*". The setting also includes the institutional and governmental aspects required to deliver health ser-vices, to address health inequalities, and to intervene for public health. Ecosystem services derives from ecological economics, which starts from the assumptions that main-stream market economics externalises the environment, and that when taken for granted, places our environment and human well-being at peril. The marriage of these concepts is rarely achieved in natural resource management, let alone public health, yet it offers considerable scope to both (Horwitz and Finlayson 2011).

The key message from this book then is that wetlands (as places of water on land, and where water shapes the land), and human health (which in its richest sense addresses the well-being of people, beyond ill-health or the absence of disease), are interconnected and to a certain extent interdependent. For multiple reasons outlined herein, there are benefits to be gained when the public sector in general, and the health sector in particular, might intervene to enhance human well-being by addressing the erosion of ecosystem services in wetlands.

References

Carter M (2015) Wetlands and health: how do urban wetlands contribute to community well-being? In: Finlayson CM, Horwitz P, Weinstein P (eds) Wetlands and human health. Springer, Dordrecht, pp 149–167

Carver S, Slaney DP, Leisnham PT, Weinstein P (2015) Healthy wetlands, healthy people: mosquito borne disease. In: Finlayson CM, Horwitz P, Weinstein P (eds) Wetlands and human health. Springer, Dordrecht, pp 95–121

Cook A, Speldewinde P (2015) Public health perspectives on water systems and ecology. In: Finlayson CM, Horwitz P, Weinstein P (eds) Wetlands and human health. Springer, Dordrecht, pp 15–30

Cunningham A (2015) Wetlands, well-being, food security and medicinal products. In: Finlayson CM, Horwitz P, Weinstein P (eds) Wetlands and human health. Springer, Dordrecht, pp 31–44

Derne B, Weinstein P, Lau CL (2015) Wetlands as sites of exposure to infectious diseases. In: Finlayson CM, Horwitz P, Weinstein P (eds) Wetlands and human health. Springer, Dordrecht, pp 45–74

Finlayson CM, Horwitz P (2015a) Wetlands as settings for human health—the benefits and the paradox. In: Finlayson CM, Horwitz P, Weinstein P (eds) Wetlands and human health. Springer, Dordrecht, pp 1–14

Finlayson CM, Horwitz P (2015b) Wetland wise use and human health—guidance for wetland managers. In: Finlayson CM, Horwitz P, Weinstein P (eds) Wetlands and human health. Springer, Dordrecht, pp 227–250

Finlayson CM, Weinstein P (2008) Wetlands, health and sustainable development—global challenges and opportunities. In: Ounsted M, Madgwick J (eds) Healthy wetlands, healthy people. Wetlands International, wageningen, The Netherlands, pp 23–40

Horwitz, P, Finlayson CM (2011). Wetlands as settings: ecosystem services and health impact assessment for wetland and water resource management. BioScience 61:678–688

Horwitz P, Roiko A (2015) Wetlands as sites of exposure to pollution and toxicants. In: Finlayson CM, Horwitz P, Weinstein P (eds) Wetlands and human health. Springer, Dordrecht, pp 75–94

Horwitz P, Finlayson M, Weinstein P (2012) Healthy wetlands, healthy people: a review of wetlands and human health interactions. Ramsar Technical Report No. 6. Secretariat of the Ramsar Convention on Wetlands, Gland, Switzerland, & The World Health Organization, Geneva, Switzerland

Horwitz P, Finlayson CM, Kumar R (2015) Interventions required to enhance human well-being by addressing the erosion of ecosystem services in wetlands. In: Finlayson CM, Horwitz P, Weinstein P (eds) Wetlands and human health. Springer, Dordrecht, pp 193–223

Jenkins A, Jupiter S (2015) Wetlands as places that help absorb the damage of natural disasters. In: Finlayson CM, Horwitz P, Weinstein P (eds) Wetlands and human health. Springer, Dordrecht, pp 169–192

McCartney M, Rebelo L-M, Sellamuttu Senaratna S (2015) Wetlands as livelihoods and contributions they make to health and well-being. In: Finlayson CM, Horwitz P, Weinstein P (eds) Wetlands and human health. Springer, Dordrecht, pp 123–148
MEA (Millennium Ecosystem Assessment) (2005) Ecosystems and Human Well-being: Synthesis. Island Press, Washington, DC
Raudesepp-Hearne C, Peterson GD, Tengö M, Bennett EM, Holland T, Benessaiah K, MacDonald GK, Pfeifer L (2010) Untangling the environmentalist's paradox: why is human well-being increasing as ecosystem services degrade? BioScience 60:576–589
WHO (World Health Organization) (1986) Ottawa charter for health promotion. World Health Organization, Geneva

Printed in the United States
By Bookmasters